ARTIFICIAL SAND FILLS IN WATER

A.A.BALKEMA / ROTTERDAM / BROOKFIELD / 1992

CUR Report 152

Colophon
Translator and editor: J. van 't Hoff
Drawings by: P. den Boer
Corrector: L. L. Kamminga
Photographs: HAM International Dredging Contractors B.V., Royal Boskalis Westminster
B.V., Lievense Engineering Consultancy B.V., Rijkswaterstaat, Van Oord ACZ B.V., Aero-
photo (Schiphol), Hylton Warner & Co. Ltd.
Printed by: W. D. Meinema B.V.

CUR and those associated with this publication have exercised all possible care in compiling
and presenting the information contained in it. This information reflects the state of the art at
the time of publication. Nevertheless the possibility that inaccuracies may occur in this
publication cannot be ruled out. Anyone wishing to use the information in it will be deemed to
do so at his or her own risk. CUR declines – also on behalf of all persons associated with this
publication – any liability whatsoever in respect of loss or damage that may arise in
consequence of such use.

Published and distributed for CUR, Gouda by
A.A. Balkema, P.O. Box 1675, 3000 BR Rotterdam, Netherlands
A.A. Balkema Publishers, Old Post Road, Brookfield, VT 05036, USA

ISBN 90 5410 138 5

PREFACE

The research committee C 56 „Artificial sand fills in water" was formed in December 1986. The aim of the research committee was to: „improve the knowledge and skills involved in the construction of artificial sand fills in water and to make this available to the industry and to Government." This would also serve to support the dredging industry's market position abroad. This aim was to be met through a publication of a manual in English. Additional practical research into construction methods was also initiated to fill in the gaps in present knowledge (technology).
It was later decided to publish the manual in Dutch as well and to include a number of case studies carried out in 1990.

The following were members of the committee C 56 involved in the preparation of the manual:

P. STRUIK, Chairman
H. VERWOERT, Secretary
W. T. BAKKER
A. L. P. ESTOURGIE
M. B. DE GROOT
J. P. F. M. JANSSEN
J. DE NEKKER
H. POSTMA
F. C. VAN ROODE
G. L. M. VAN DER SCHRIECK
J. C. WINTERWERP
J. VAN 'T HOFF, Rapporteur and Editor
G. J. H. VERGEER, Coordinator
Prof. Dr. J. F. AGEMA, Mentor

L. VERSTOEP was chairman of the research committee until November 1988, when he was succeeded by P. STRUIK. J. P. F. M. JANSSEN served as Secretary of the committee until November 1988, when he was succeeded by H. VERWOERT. H. N. C. BREUSERS was a member of the committee until September 1988, when his place was taken by J. C. WINTERWERP. G. L. M. VAN DER SCHRIECK and A. L. P. ESTOURGIE have been members of the committee since 1989. J. VAN 'T HOFF has been rapporteur since September 1989.
As well as members of the research committee the following persons have contributed to the manual: E. G. J. VAN KUIJK of the Rotterdam Municipality, D. R. MASTBERGEN of Delft Hydraulics, E. V. MEIJER of the Delft Technical University and M. ZWIERS of Aveco.

3

The final text was edited and translated by J. VAN 'T HOFF, consulting engineer. The English translation was reviewed by Dr. A. R. CLARK of Rendel Geotechnics.

CUR thanks the Road and Hydraulic Engineering Division of the Netherlands Department of Public Works (Rijkswaterstaat), the Department of Economic Affairs and the dredging companies Royal Boskalis Westminster B.V. and the HAM International Dredging Contractors B.V. for the financial contributions which helped to make this manual possible.

CUR also thanks Delft Hydraulics and Delft Geotechnics for their contributions to the theoretical section of the manual and the dredging companies Royal Boskalis Westminster B.V., HAM International Dredging Contractors B.V., the former Volker Stevin Dredging B.V. and the Public Works Rotterdam Harbor Engineering Division for their contributions to a number of chapters on practical issues and to the case studies.

July 1992   The Executive Committee of the CUR

CONTENTS

THEORY

5

PRACTICE

LIST OF SYMBOLS

A	cross-section of pipeline	m^2
A	effective current exposed area	m^2
B	bulking	
B	spreading width of mixture	m
B_{max}	upper limit of spreading width of mixture	m
B_{min}	lower limit of spreading width of mixture	m
b	width of plane shaped jet and fill area	m
b_0	width of opening for plane shaped jets	m
C_s	pressure coefficient	
c	volume concentration	
c_b	volume concentration just above sea bed	
c_v	volume concentration	
c_t	transport concentration	
c_g	apparent concentration	
c_0	volume concentration at point of discharge	
D	grain diameter	μm
D_r	relative density	
D_{re}	relative density based on void ratio	
D_0	reference value of grain diameter	μm
D_{10}	particle size (90 % by weight exceeded in size)	μm
D_{15}	particle size (85 % by weight exceeded in size)	μm
D_{50}	median particle size (50 % by weight exceeded in size)	μm
D_{60}	particle size (40 % by weight exceeded in size)	μm
D_{80}	particle size (20 % by weight exceeded in size)	μm
D_{rn}	relative density based on porosity	
d	diameter of round jet	m
d_b	diameter of round jet just above sea bed	m
d_0	diameter of nozzle	m
e	void ratio	
e_{max}	maximum void ratio	
e_{min}	minimum void ratio	
F	force exerted by wind or current on body	N
f_t	transport factor	
f_0	Darcy-Weisbach friction coefficient	
g	acceleration of gravity	m/s^2
H	vertical height of nozzle from sea bed (under water)	m
H_s	significant wave height	m
$H_{s\ all}$	maximum allowable significant wave height	m

14

$H*$	soil mechanics sedimentation parameter	
HS	dimensionless parameter in equation (7)	
h	slope height	m
h_{cr}	critical slope height	m
h_p	vertical height of nozzle above water level	m
i	slope gradient	
k	permeability coefficient	m/s
L	sedimentation length	m
L_m	sedimentation length in model	m
L_p	sedimentation length in prototype	m
$L*$	hydraulic sedimentation parameter	
$N*$	soil mechanics sedimentation parameter	
n	porosity	
n_{dam}	porosity in sand dam	
n_{crit}	critical porosity	
n_{max}	maximum porosity	
n_{min}	minimum porosity	
P	sand production including pores	m^3/s
p	specific sand production	m^2/s
Q	total flow rate	m^3/s
Q_g	sand production including pores	m^3/s
Q_s	production of solids	m^3/s
Q_m	production of mixture	m^3/s
q	specific (mixture) flow rate	m^2/s
q_c	cone resistance	MPa
q_0	specific mixture flow rate at point of discharge	m^2/s
r	distance to jet centre	m
r_∞	crater radius or half the crater width	m
$r_{\infty s}$	crater radius for clear water	m
s	specific sand production (solids)	kg/ms
s	specific sand transport (solids)	kg/ms
T_s	significant wave period	s
t_i	process periods	s
U	mixture velocity in centre of jet	m/s
U_a	current velocity of surrounding water	m/s
U_b	mixture velocity in centre of jet just above sea bed	m/s
U_c	critical mixture velocity	m/s
U_m	(mixture) velocity on plate or in model	m/s
U_p	(mixture) velocity in prototype	m/s
U_0	(mixture) discharge velocity	m/s
u_*	shear velocity	m/s
u_{*cr}	critical shear velocity	m/s

15

V	volume	m^3
V_s	volume of solids	m^3
V_m	volume of mixture	m^3
v_m	velocity of sand-water mixture in pipeline	m/s
v_s	velocity of solids in pipeline	m/s
W_0	fall velocity of individual grains	m/s
W_{50}	median fall velocity	m/s
x	horizontal distance from centre of jet to point of discharge	m
x_b	horizontal displacement of jet by current	m
y	erosion depth for sand-water mixture	m
y_s	depth of erosion crater for clear water	m
z	vertical distance to point of discharge	m
α	slope angle	°
Δ	relative density of sand particles	
φ_0	angle of incidence of mixture flow	rad
ϱ_s	density of solids	kg/m^3
ϱ_m	density of sand-water mixture	kg/m^3
ϱ_w	density of water	kg/m^3
ϱ_0	density of sand-water mixture flow at point of discharge	kg/m^3
σ	effective stress	Pa
τ	shear stress	Pa

ABBREVIATIONS

ASTM	Acoustic Sand Transport meter
CUR	Centre for Civil Engineering Research and Codes
CCM	Conductive Concentration meter
CD	Chart Datum
DG	Delft Geotechnics
DH	Delft Hydraulics
EMF	Electro Magnetic Flow meter
EMV	Electro Magnetic Velocity meter
GPS	Global Position System
KIVI	Royal Institue of Engineers
M.S.L.	Mean Sea Level
ROV	Remotely Operated Vehicle
RWS	Rijkswaterstaat
SPT	Standard Penetration Test
VAT	Visual Accumulation Tube
ZEF	Zone of Established Flow
ZFE	Zone of Flow Establishment

CHAPTER 1

INTRODUCTION

1.1 General

Across the whole world projects of various sizes are executed in which the construction of a sand fill in water forms a part. These projects, among others, include the building of dams for the purpose of wave reduction or for the benefit of (tidal) closures, the formation of artificial islands or coastal extensions, the covering of pipelines and the foundation of submerged tunnels and caissons.

In the Netherlands experience has been gathered on a great number of these types of project especially in the large ports, pipeline covering in the North Sea and within the framework of the Deltaworks for the defence against storm surges of the southwestern part of the Netherlands.

Additionally Dutch contractors, research institutes and consultancy firms have been involved in the research, design or construction of a great number of projects abroad. These varied experiences are based on in depth research; especially within the framework of the preparations of a number of Delta-closures to be executed with sand. These involved research into the literature and theory development as well as experiments on a laboratory scale and prototype site observations. Moreover the execution of some projects has been extensively monitored including specific measuring programmes to broaden and to deepen the knowledge.

After the completion of the Deltaworks it seemed important not to leave this knowledge scattered over a great number of study reports and design and evaluation notes, but rather to achieve a userfriendly compilation of the data and experience. As a result of this aim this manual has been produced.

1.2 Use and objectives

After a general introduction to the area of knowledge this manual is separated into two main parts, namely theory and practice.

The theoretical part starts with a description of the processes which occur during construction with sand in water so that a review of the relevant phenomena is obtained. For this the sequence of the sand, sand-water mixture movement is considered. Subsequently the developed theory is described for each process, at which time the relevant parameters and the relationship between them are addressed. The theoretical part is concluded with a summary of the theory for practical use, in which the most important relationships are brought together.

The practical part deals firstly and separately with the aspects which are required for the design and the execution of the project. This concerns the field boundary conditions, the project to be constructed, the available work methods and the equipment for the winning and the placement of sand. Furthermore the volume measurements and the quality assurance play their role. For each project to be constructed a combination of the above mentioned factors applies. The achievement of a good design relies mainly on the optimal and intensive search of construction shape, working methods and equipment selection for the identified boundary conditions and established functional requirements.

The design is a complicated and strongly iterative process of inventive and flexible anticipation of the presented circumstances with the available possibilities. This already complicated process of combining and integration is difficult to describe in an abstract form. For that reason the choice has been made to illustrate this with a number of case histories. In order to indicate that this does not demand a rigid prescription, but a specific approach per project, the number of case histories is considerable. At the same time each of the described processes in the case histories is related back to the relevant theory.

Finally the practical part is concluded with a worked out practical example of a case history, in which all aspects of a specific project are dealt with. The emphasis here is on the translation of the theory into practice.

Throughout this manual an attempt has been made to present the available knowledge in the field of construction of sand fills in a user friendly manner and make it widely known. The manual intends to support the design and execution of future works. It is primarily aimed at technicians directly involved in the preparation or execution of the construction of sand fills under water. However, owing to the format of the manual with a separate theoretical and technical part, the possibility for use is further increased.

For the more interested reader, such as technical students or non-technicians involved in such projects, the general project description at the beginning of each theoretical section and the practical part may provide useful insight.

Finally the theoretical part provides a survey of the present state of the art in this field which, added with a literature overview, may be a useful guide for researchers starting in this field.

1.3 Continuation

Apart from the publication of this manual the committee has made an inventory of gaps in the present knowledge, on which basis proposals are made for further research. This especially concerns the influence of waves on the slope in the tidal zone, the control of (deep) underwater processes and the compaction of the fill area.

Furthermore the wider distribution of this knowledge may possibly provide an extra stimulus to further development of supplementary knowledge. For this reason a questionnaire is included (see Appendix G), through which the user of this manual is kindly requested to deliver available new information, especially on practical experience of completed projects. With this and further research yet to be carried out, a new manual or a supplement may be prepared.

At the same time the reader is requested to mention possible omissions or inaccuracies in the manual, preferably well documented, so that these may be included in a next edition or supplement.

theory

CHAPTER 2

PROCESS PHENOMENA OF BUILDING WITH SAND IN WATER

2.1 Summary

In this summary the different phases are described of taking up, transporting and placement of sand for the construction of sand fills in water.
Figure 1 shows these phases schematically. In this Figure distinction is made between
the following elementary methods of placing sand:

- single point discharging above water;
- single point discharging under water;
- curtain-like discharging;
- lump-like discharging.

The first method can also be described with the term „horizontal" placement and the
latter three methods can all be described with the term „vertical" placement. Despite
this distinction, the process to which the sand is subjected while being placed can
almost always be described in the same manner. The indicated phasing is especially
applicable to tidal closures using sand, beach nourishment, the construction of a submerged sand berm etc. Also for other applications of building with sand in water the
different stages are almost identical: such as the single point discharging under caissons
or the pipeline placement in enclosed basins.
Table 1 gives an overview of the different phases as indicated in Figure 1 in the
sequence of taking up, transporting and placement. The paragraphs in which the
different phases are described are also indicated in this Table.
After winning (phase I), the sand is transported to the fill area (phase II) by hopper
(dredge) or via a pipeline. Subject to the means of transport, the sand is either bottom
discharged or pipeline placed (phase III). The sand-water mixture jet makes craters

Table 1. Overview of the different phases as indicated in Figure 1 for the construction of sand
fills in water.

phase	description	paragraph
I	sand winning	2.2
II	sand transport	–
III	formation of sand water jet	2.3
IV	formation of a crater	2.4
V	flowing of sand-water mixture on slope above water	2.5
VI	flowing of sand-water mixture on submerged slope	2.6
VII	loss of sand under water	2.7
VIII	sedimentation and formation of slope	2.8

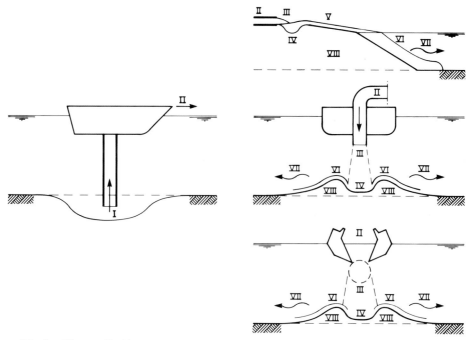

Fig. 1.   Phases of taking up, transporting and elementary methods of sand placement.

Rainbowing in the Beaufort Sea.

(phase IV) in the sand body under construction. A sand-water mixture flows across the edge of these craters. This overflow can occur above water (phase V) as well as under water (phase VI). Underwater loss of sand (phase VII) to the surrounding water can occur. Sand in the mixture flow will settle rapidly to form a slope (phase VIII).

This manual mainly describes the processes to which the sand is subjected during hydraulic placement. Sand winning is only dealt with summarily, while the transporting of sand is not dealt with any further as many handbooks are available on this subject.

2.2 Slope formation in the borrow pit

To a great extent, the size of a borrow pit is determined by the slope gradient resulting from the winning of sand. The ultimate slope gradient is also important for the minimum distance between the borrow pit and the discharge area.

The slopes in a borrow pit are mainly determined by the characteristics of the dredged sand. This sand originates in the first place from the direct surrounding of the suction mouth. Depending on the manner in which the slope is formed, the sand can also originate from soil layers situated higher up on the slope.

When the suction takes place at the toe of a relatively steep slope, sand situated higher up on the slope will flow to the suction mouth of the dredger. This can occur by way of thin grain flows or by way of a hyperconcentrated sand-water mixture flow. In that case a regular process of slope formation and slope retreat develops. However, during a slip failure or a flow slide an irregular process of slope formation will develop, in which very large volumes of sand will flow to the suction mouth.

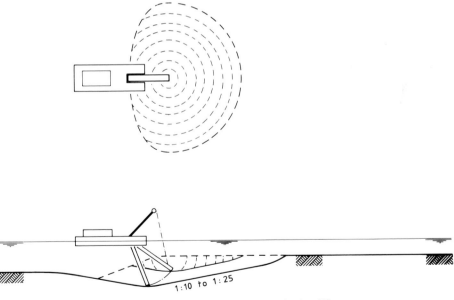

Fig. 2.   Flow slide during sand winning [1].

27

A slip failure develops when a very steep slope is undermined at the toe. Locally a volume of sand will then slide off. The resulting slope (1 : 1.5 to 1 : 2) depends on the type of sand and on the hydrostatic pressure. Slip failures can also occur after stopping the suction process.

Flow slides do occur in loose sand slopes. The sand body liquefies and flows out over a distance of 100 m or more. This phenomenon is of a greater order than slip failure. Slopes resulting from flow slides are often flatter (1 : 10 to 1 : 25) than natural slopes (see Fig. 2).

The dredged sand can be transported to the sand fill area by hopper barge or via a pipe-line. The discharge process will be discussed in 2.3.

2.3 Falling of a sand-water mixture through water

Entrainment

During single point discharging under water and curtain-like discharging, sand falls through the water column after leaving the pipeline or hopper. Under water the sand-water mixture still exists. However, the mixture absorbs water from the surrounding fluid (entrainment). This changes the jet geometry. For example during single point discharging under water, the jet diameter or the jet width increases with depth in a ratio of approximately 1 : 5 ($=$ 2 times 1 : 10). Hence with every 5 metres of depth increase the diameter or width increases by 1 metre (see the hatched line in Fig. 3). The increase of

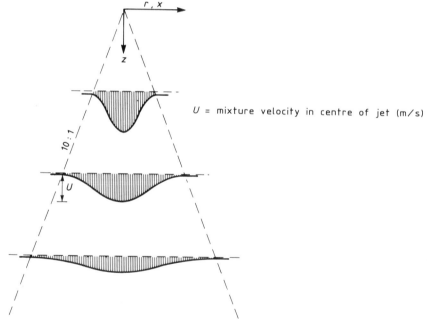

U = mixture velocity in centre of jet (m/s)

Fig. 3. Velocity profiles at different depths during single point discharging under water and during curtain-like discharging.

28

water content results in an increase of the mixture-flow and a decrease of the sand concentration and mixture velocity. Part of the sand can be diffused into the surrounding water and therefore disappears from the mixture-flow. This may result in a loss of sand (see 2.7). In practice it is not possible to draw a distinct line between the jet and the surrounding fluid during single point discharging under water or curtain-like discharging.

Acceleration and deceleration
After leaving the pipeline or hopper the sand-water mixture will accelerate or decelerate. The acceleration is caused by the difference in density with the surrounding water and by gravity. The deceleration is a result of mixing and friction with the surrounding water. The resulting effect of acceleration or deceleration depends on the dominance of the above effects. With single point discharging under water (vertical downward) the deceleration factor often dominates and the velocity will only decrease. After leaving the pipeline, the velocity profile will gradually taper off into a more or less Gaussian shape (see Fig. 3).
During curtain-like discharging and lump-like discharging the initial velocity equals zero. Here a distinct acceleration occurs directly after leaving the hopper. The velocity increases to a maximum and then decreases again. During lump-like discharging the sand falls in first instance like a cylindrical shaped mass, which transforms into a drop shape while falling (see Fig. 4).

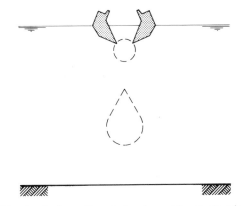

Fig. 4.   Deformation during lump-like discharging.

Diffraction
Diffraction of a sand-water mixture is another important phenomenon during the falling of a sand-water mixture through water. Diffraction is defined as the horizontal displacement of sand in relation to the discharge point. This can, among others, be caused by a cross current but also when the angle of incidence of the sand-water mixture is not perpendicular to the water surface. For horizontal hydraulic placement

of sand this angle of incidence depends also on the height of the discharge pipe above the water surface (see Fig. 5).

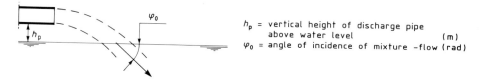

h_p = vertical height of discharge pipe
      above water level   (m)
φ_0 = angle of incidence of mixture –flow (rad)

Fig. 5.   Angle of incidence for horizontal placement of sand.

The diffraction of a mixture flow caused by a cross current is shown in Figure 6. The extent of this bearing off is mainly determined by current velocities and differences in density.

Diffraction will partly cause the sand-water mixture to segregate. The heavier particles will settle sooner and consequently end up closer to the discharge point. The cross current may lead to more entrainment and to possible sand losses.

Finally an increase in water depth will result in a proportionally larger bearing off. This

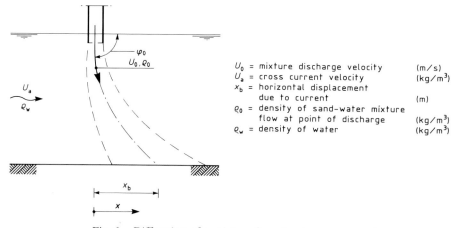

U_0 = mixture discharge velocity   (m/s)
U_a = cross current velocity   (kg/m³)
x_b = horizontal displacement
     due to current   (m)
ϱ_0 = density of sand-water mixture
     flow at point of discharge (kg/m³)
ϱ_w = density of water   (kg/m³)

Fig. 6.   Diffraction of a mixture flow due to cross current.

is caused by a decrease in vertical velocity with depth and an increasing jet diameter. Eventually the sand-water mixture reaches the bottom with a certain velocity whereby a crater will be created. This phenomenon is dealt with in 2.4.

2.4 Formation of craters

The sand-water mixture jet impinging on the bottom creates a crater with a turbulent mixture flow inside. This scour hole is the result of both erosion and sedimentation (see

Fig. 7). However, when jetting occurs at the same location for an extended period, erosion and sedimentation are in balance with each other and the hole reaches its equilibrium shape. The mixture flows across the crater edge which operates as a spillway.

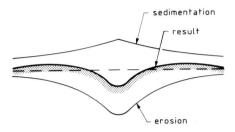

Fig. 7.   Erosion and sedimentation in a crater.

Craters can occur both below water and above water (see the Figures 8 and 9). Submerged craters are more important because they determine to a great extent the ultimate shape of the sand body.

Craters are not only created by single point discharging under water or curtain-like discharging but also by lump-like discharging. An example of the latter is given in Figure 10. In this specific case 9 to 12 hopper loads were discharged on one spot. The depicted cross section applies to the discharge location. Within the channel deepening occurs whereas next to the channel the bed will be raised. Only part of the sand will be incorporated in raising of the bed next to the discharge location in the longitudinal direction of the channel.

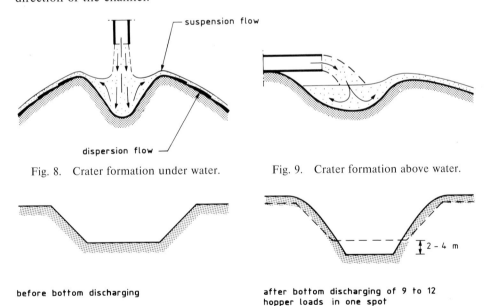

Fig. 8.   Crater formation under water.   Fig. 9.   Crater formation above water.

before bottom discharging   after bottom discharging of 9 to 12
                                             hopper loads in one spot

Fig. 10.   Crater formation for lump-like discharging in a channel.

In Figure 11 the crater diameter and the sedimentation area are defined for hydraulic placement of sand under water. The order of magnitude of a crater diameter $(2r_\infty)$ is approximately 10 to 50 m for mixture-flows of approximately 1 to 10 m^3/s (this includes curtain-like and lump-like discharging). The sedimentation area L_h is always larger. The sand-water mixture flows over the crater edge and continues down the crater slope. The phenomena which occur during the flowing of a sand-water mixture above water are described in 2.5, and under water in 2.6.

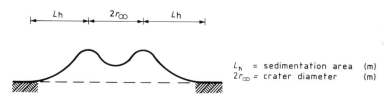

L_h = sedimentation area (m)
$2r_\infty$ = crater diameter (m)

Fig. 11.   Crater diameter and sedimentation area.

2.5   Flowing of a sand-water mixture above water

After the mixture flow has left the crater the flow will generally accelerate or decelerate. The term mixture flow still applies. Acceleration occurs for slopes which are too steep with erosion as a concequence. Deceleration occurs for slopes which are too flat with sedimentation as a concequence. On the fill area above water an equilibrium slope develops when sedimentation and erosion are in equilibrium with each other. This slope is the reference value for the following observations.

With time a fill area increases when sand is supplied. The temporary slope is therefore somewhat flatter than the equilibrium slope. The acceleration or deceleration is modified and dampened as a result of the adjustment of the slope. A slope will adjust to changes in water level. For instance when the water level is lowered an initial above-water slope develops which is too steep, and only later after erosion, a slope develops which is equal to the equilibrium slope (see Fig. 12).

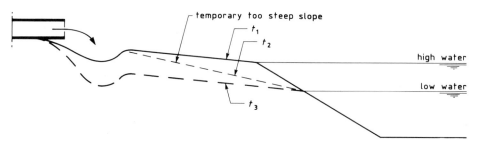

Fig. 12.   Adjustment of the slope inclination to the lowering of the water level.

32

Abovewater fill area with well distributed mixture flow.

When taking into account the fill area above water as a whole, changes only occur slowly because the mixture flows and concentrations will alter only slowly between the discharge pipe and the water line. Locally however, the flow and sand transport may vary considerably. On the abovewater portion of the fill area terraces with hydraulic jumps may develop. Also channels may develop which can cause the mixture flow to meander in the direction of the water line.

Hydraulic jumps and terrace formation
Under certain conditions a pattern of alternating steep and gentle slopes develops with hydraulic jumps directly after the steep slopes. These hydraulic jumps in a sand-water mixture, or „mixture-jumps", occur at regular distances of the order of some meters. On the abovewater fill area three types of features can be distinguished: cascade, hydraulic jump and terrace (see Fig. 13).

Through erosion at the cascades and sedimentation on the terraces, the pattern of cascades and terraces moves upstream with a velocity of approximately 0.01 to 0.03 m/s. A terrace shaped fill area occurs particularly when the sand-water mixture flow is distributed across a large width resulting in small specific flow rates ($q = 0.01$ to 0.3 m^2/s). This occurs particularly at high water levels (hence at short fill length).

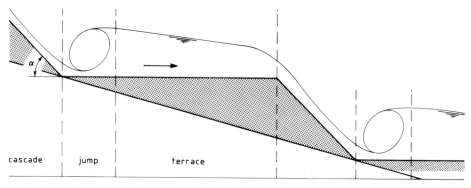

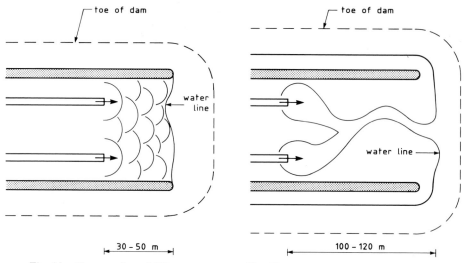

cascade | jump | terrace

Fig. 13.   Fill area above water with hydraulic jumps.

In practice, a terrace shaped fill area above water of 30 to 50 m often occurs (see Fig. 14).

Channel formation

Besides terrace shaped fill areas, also fill areas with channels can develop in which the sand-water mixture concentrates (see Fig. 15).

Fig. 14.   Terrace shaped fill area.

Fig. 15.   Abovewater fill area with channel formation.

Also in the channels cascades and terraces will sometimes develop. In that case the lengths (approximately 10 m) and the heights of the cascades (approximately decimeters to metres) are relatively large.

Channel formation is best developed on slopes steeper than the equilibrium slope. The specific flow rate is large because of the concentration of the flow in the channel and varies between 0.3 m²/s and 1.0 m²/s. Because of this large specific flow rate, the average slope of the channel is flatter than the slope of the terrace shaped fill area. This flat slope also results in a flat slope of the fill area above water and depending on the height of the discharge pipe above water level, sometimes reaches lengths of more than 100 m. Towards the water level the concentration changes very little.

Meandering
Together with channel formation, meandering of the sand-water mixture may occur in the direction of the water level. Because the meandering flow can move freely across the fill area above water, bunds which may be present, may erode. This erosive action is greater than in case of terrace formation, due to higher flow velocities.

2.6 Flowing of a sand-water mixture under water

The flowing of a mixture across an underwater fill area shows great similarity with the flowing of a mixture on a fill area above water as described previously.

The difference with flowing above water is firstly the occurrence of steeper slopes because the driving force is smaller as a result of the effect of buoyancy and secondly because of the mixing of water from the surroundings with the density current. This may lead to a change in sand concentration in the density current and can lead to loss of sand from the works.

The sand-water mixtures which end up in the water, or flow across the edge of the crater, continue to flow as more or less independent density currents. Just like the fill area above water, changes of the mixture flow and concentration are slow and occur across a relatively large distance. However, this distance is shorter than above water because the slopes are steeper and because the current velocities are smaller. Hence under water the density current maintains itself less well than above water.

When a slope of the underwater fill area is steeper or flatter than the equilibrium slope the flow will accelerate or decelerate respectively just as on the fill area above water. The equilibrium slope under water is more often steeper than above water. Also under water the slope inclination adjusts itself to the equilibrium value.

Locally strong deviations of the equilibrium slope can occur because of hydraulic jumps and terraces. In this case the flow alternately accelerates and decelerates. If the mixture velocity is beyond a certain critical value, an internal hydraulic jump occurs when the flow decelerates (see Fig. 16). The heights of the cascades are greater and the lengths of the terraces are smaller in comparison with the fill area above water. Generally speaking, the pattern of terraces and cascades under water is less surveyable and recognizable than above water. The result is an underwater fill area where locally great differences in the slopes of the bottom can be distinguished.

Under water the sand-water mixture flow meanders in the same manner as above water. This appears from frequent soundings.

After the initial settlement the sand can start moving again because of a flow slide. This flow has a high concentration and moves slowly.

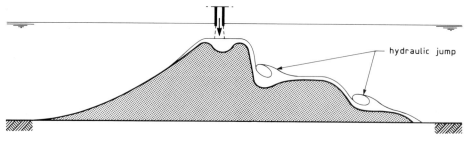

Fig. 16. Hydraulic jumps under water.

2.7 Loss of sand under water

During the falling of sand through water and during the flowing of the density current on the bottom under water, loss of sand will occur. In both cases the same process applies namely the exchange of a sand-water mixture with the water surrounding the jet and the water above the density current. Sand and water may be exchanged between the sand-water mixture and the ambient water.

This departing sand does not necessarily have to imply a loss from the sand body. The losses depend on the definition of the boundaries within which the sand can be regarded to form part of the sand body and outside which the settled sand can be regarded as losses (see Fig. 17).

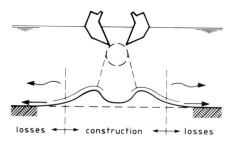

Fig. 17. Definition of sand losses.

Sand lost to the ambient water may still settle within the boundaries of the sand body to be constructed.

However the mixture-flow may also transport sand outside the boundaries of the sand body to be constructed and in this way contribute to the losses.

Finally erosion of previously settled sand may contribute to the total sand losses.

2.8 Sedimentation and slope formation

Sedimentation

As soon as a mixture flows across a slope which is flatter than the equilibrium slope, sedimentation begins. Sedimentation needs time and needs even more time when the grain diameter is smaller, the mixture flow has a greater layer thickness and the concentration is higher. Consequently sedimentation takes place over a certain distance. This distance is greater as the time for settling is longer and the current velocity is higher (see the Chapters 6, 7 and 8). The packing of sand placed under water is usually loose.

The construction of a sand body under water may be combined with flow slides or the development of quicksand. The circumstances for which these phenomena can occur will be discussed in more detail in this section as well as the slopes and shapes of the fill area which can result from the sedimentation of sand. Finally the phenomena of sieving and erosion will be discussed briefly.

Flow slides

A small disruption of loose sand under water may cause the sand to liquefy and to flow away. This then is called a flow slide (see Fig. 18).

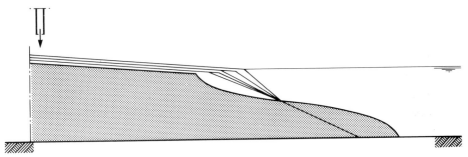

Fig. 18.   Flattening of the underwater slope as a result of a flow slide.

The following conditions and circumstances are of importance with respect to flow slide:

– steep slopes, steeper than 1:3 to 1:7;
– sand of low permeability because of the presence of fine sand and silt;
– a porosity greater than the critical value.

The flow slide compacts the sand to a certain extent and a flat slope develops. However this does not necessarily mean that after the first flow slide the packing of the sand is such that another flow slide cannot occur. Because of this, various flow slides can occur at one spot during the formation of a slope. This can especially be the case when the sand is less than medium grained ($D_{50} < 250$ μm) and when large portions of the slope are steeper than 1:3 to 1:7.

The depth of a flow slide can be several tens of meters and the time involved is of the order of some minutes. The slopes which develop as a result of a flow slide are generally two to three times flatter than the equilibrium slope after first sedimentation.

Quicksand

Related to flow slides is the occurrence of quicksand. Just like the sand during flow slides, quicksand is liquefied sand. However, contrary to flow slides quicksand does not flow but behaves as a heavy fluid when surcharged. Spots of quicksand occur on the above-water fill area(s), making this (these) difficult to access. Low permeability promotes the occurrence of quicksand. Because, among others, the presence of a high groundwater level and in general a low local permeability, quicksand occurs more often close to the waterline.

Slope gradients
Slopes are in the first place determined by the grain size diameter. Coarser grains result in steeper slopes.
Above water the specific mixture flow rate is of importance for the slope gradient. A larger specific mixture flow rate results in a flatter slope.
Under water the specific sand transport has a strong influence. A larger specific sand transport rate results in a flatter slope.
Finally the method of hydraulic placement is of importance (see the Chapters 7 and 8). Figure 19 gives some practical values of underwater slopes for single point discharging above water and for grain size diameters (D_{50}) between 150 and 250 µm.

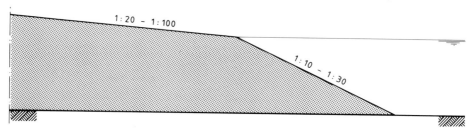

Fig. 19.   Slope gradients for single point discharging above water.

For single point discharging above water, the underwater front slope differs sometimes from the side slope (see Fig. 20). Slopes which are steeper than the side slopes as well as vice versa have been observed in practice.

Fill area formations
If there are no limitations by way of embankments, quays or walls and there are no out-side influences, the natural shape which develops during the hydraulic placement of sand in elevation will be a delta or cone. For the horizontal single point discharging, the cone-shape above water is regulated by outside influences like bunds and bulldozers. Under water, the cone-shape can often be clearly distinguished.
The cone-shape can at best be observed during vertical hydraulic placement (see Fig. 21). After extended single point discharging in one location, a conical crater develops. Comparable to the meandering of the flow on an abovewater fill area is the rotation of the flow on a submerged fill area during vertical single point discharging. Flowing out of a crater occurs here in a hyperconcentrated flow. This flow is not fixed but moves along the edge of the crater (see Fig. 22). A rotating flow also occurs during the single point discharging under caissons (see the relevant case history in Chapter 20). This flow enables the complete filling of the space underneath a caisson with sand.

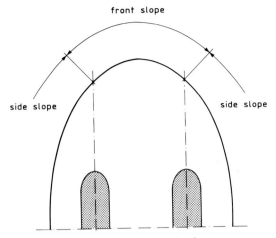

Fig. 20.   Front and side slopes.

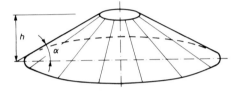

Fig. 21.   Cone-shape during vertical placement.

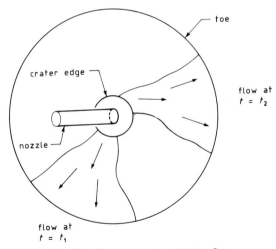

Fig. 22.   Top view of a rotating flow.

Sorting

During sedimentation sand will be sorted. This is the separation of sand according to fall velocity or grain size. On average the coarser fractions settle sooner. The finer fractions settle more slowly and consequently will end up at a greater distance. Sorting

39

occurs below as well as above water level (see Fig. 23). The condition for sorting is sand with a broad grading and not too high sand concentration.

Cross currents may transport the finer fractions outside the planned sedimentation area (see also 2.3).

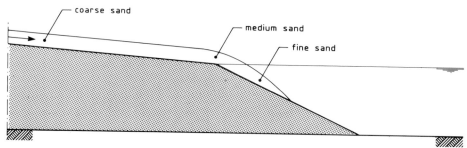

Fig. 23.   Sorting below and above water level.

Erosion

The last phenomena to be discussed is erosion by currents and waves. Sand will be stirred by waves and currents after which the current transports this re-suspended sand. Sand can be transported large distances outside the area of the sand body under construction. Currents will erode the underwater slope over the full height, whereas waves will erode the sand especially near the water level. In the case of a tidal current where the water level varies, the erosion by waves will cover a greater area.

A varying water level will strongly influence the flow pattern on the fill area above water, consequently erosion also occurs above water (see 2.5).

How an underwater mound and side slopes of a sand fill can erode is shown in Figure 24. The erosion velocity strongly depends on the initial shape of the mound. A portion of the eroding sand will for example settle directly in an adjacent crater. The erosion velocity will then decrease.

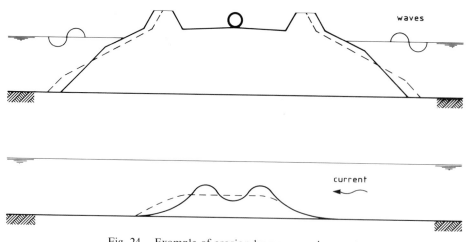

Fig. 24.   Example of erosion by waves and current.

CHAPTER 3

SOIL INVESTIGATION IN THE BORROW AREA

3.1 Need for soil investigations

Soil investigations in the borrow area or potential borrow areas can be of interest for three reasons:

- determination on the suitability of sand as a building material;
- determination on the winning method and the expected production;
- determination on the risk of loss of stability of adjacent structures as a result of the winning.

Investigation of the potential borrow area can also be necessary in relation to undesired morphological consequences of the sand winning or other undesired consequences for the environment. These will not be discussed here (see further Chapter 14).

The suitability of sand as a building material is mainly determined by the grain size distribution: the D_{50} or D_{15}, the silt content (2 µm $< D <$ 63 µm) and the clay content ($D < 2$ µm). Furthermore the presence of lime can influence the suitability.

Besides these aspects other soil conditions are also of importance for the sand winning, the selection of the sand winning method and the expected production, namely: the permeability, the packing density (especially if it concerns rather fine sand) and the layer structure (unsuitable layers, depth, thickness and extent of retrievable layers and nature of unsuitable overburden). For example if one considers the winning of sand from underneath a layer of clay without completely removing this layer, then the strength characteristics of this layer are of importance.

The risk of loss of stability of adjacent structures is only of importance if the distance from the borrow pit is relatively small. One has always to make sure that a safe distance is maintained between the structure and the top of the slope of the borrow pit (see Fig. 25). The location of the top of the slope is determined by the slope gradient which develops in the borrow pit. If no special measures are taken during the sand winning, the developed slope gradient will be the result of slip failures and flow slides.

In the event of slip failures the slope of the sand will not be much smaller than the natural slope (see Fig. 25). Extra information of the soil is required only if within or directly underneath the borrow pit soft clay layers or peat occur, resulting in flatter slopes or unacceptable deformations (see Fig. 26). In case of possible flow slides (see Fig. 27 and also Chapter 9) information on the grain size distribution of sand and the packing density is essential.

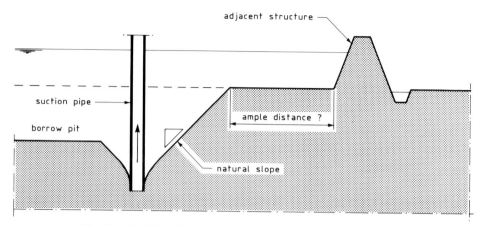

Fig. 25. Stability of adjacent structure for slope with dense sand.

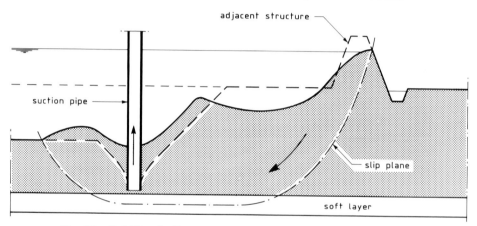

Fig. 26. Stability of adjacent structure, slip failure caused by soft layer.

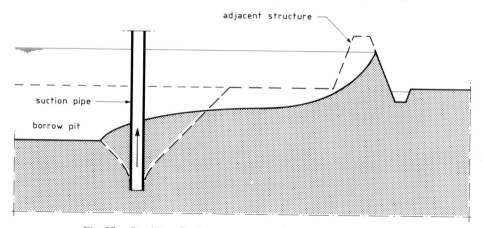

Fig. 27. Stability of adjacent structure, flow slide in loose sand.

42

3.2 Review of investigation methods

Geological and historical research
On the basis of geological and historical research the extent and nature of the soil investigation can be established. In a later stage the geological knowledge can be of use for the interpretation of the results and to assist in the decision on the need for a possibly more detailed soil investigation.
Geology enables insight with respect to:

- the type of deposits;
- the thickness and extent of the different layers;
- local variations caused by previous rivers, channels etc.;
- the question whether the present soil was surcharged in the past by layers which were later eroded again.

Historical documents may provide an impression on the probability of wrecks, submerged settlements and the like.

Taking of surface samples (with grab-sampler)
When searching for a sand winning area it is often useful to investigate first the surface of the river and sea bottom on the presence of sand.

Borings and laboratory tests
A soil investigation is hardly conceivable without at least some bore holes. There are different types of borings. The more simple types are aimed to determine the layer structure of the subsoil. The more complicated types of borings are also aimed to obtain undisturbed samples for the investigation of strength and stress parameters. Often a more simple type of boring will suffice here, for instance cable percussion borings (see Fig. 28) or vibration-borings (see Fig. 29). Wash-borings can also be useful. In Figure 30 the principle of a wash-boring is shown for wash-borings on land. When drilling from a pontoon or a vessel, the soil-water mixture is transported to the deck through a casing pipe. When applying wash-borings one has to realize that fine particles can possibly be washed away. For that reason information on the presence of fine material is essential. Details of site investigation methods are given in the „Code of practise for site investigations" issued by the British Standards Institution (BS 5930, 1981).
In all these cases the borings have to be combined with some kind of laboratory investigation on the obtained samples in order to classify the soil, such as for instance the determination of grain size distributions, Atterberg limits, densities, water content, lime content etc.
If extensive clay or peat layers occur, some tests should be carried out to determine the strength (especially undrained) for determination of the dredgeability and the stability of the slopes. Tests such as vane tests in the bore hole, or laboratory tests such as torvane tests on samples, unconfined compression tests and also some direct shear tests, triaxial- or cell tests are appropriate. For the laboratory tests undisturbed samples are required.

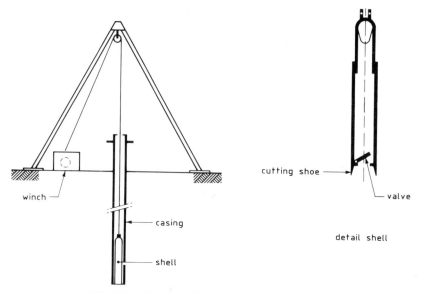

cutting shoe

winch

valve

casing

detail shell

shell

Fig. 28. Principle of a cable percussion boring.

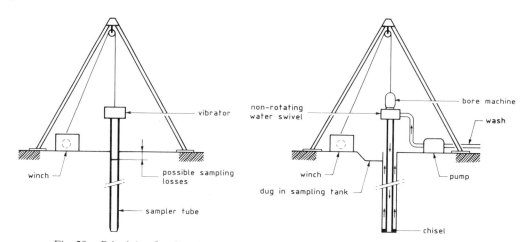

vibrator

non-rotating
water swivel

bore machine

wash

winch

possible sampling
losses

winch

pump

dug in sampling tank

sampler tube

chisel

Fig. 29. Principle of a vibration-boring.　　　　Fig. 30. Principle of a wash-boring.

Soil penetration tests (SPT's)

If there is a question of fine sand, knowledge on the permeability and the deformation characteristics (which are a function of the packing density) are required in order to predict the dredging production and the probability of flow slides. In the latter case also horizontal soil stresses are important. Borings and laboratory tests give information on the permeability but not on the packing density and the horizontal soil stress.

44

For that reason borings can be extended with soil penetration tests (SPT's) in the bore hole (see Fig. 31). These provide an indication on in situ relative density and in general terms information on horizontal soil stress. With soil penetration tests the number of times are counted that a standard weight must be dropped a standard distance in order to penetrate the splitsampler 0.3 m. The soil retrieved with the sampler can be used for classification tests.

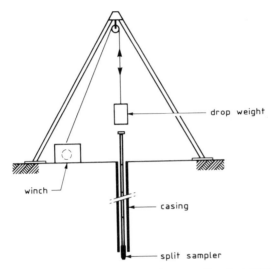

Fig. 31.   Principle of a standard penetration test (SPT).

Cone penetration tests (CPT's)
A better indication of the in situ characteristics can be obtained by means of static cone penetration tests. When the local adhesion and preferably also the local hydrostatic pressure (piezo-cone) are measured with these tests, cone penetration tests can replace many borings. It also happens that from such soundings the layer structure and the soil type can be derived, provided that in the area adjacent to a sounding a couple of borings have been carried out for calibration.
CPT's require much less time than borings. On land and in sheltered water they are often also cheaper. However this is often not the case in open sea because CPT's require special arrangements to provide the reactive force for the penetration.

Pressiometer and/or dilatometer
In a better way than with the CPT, information on the horizontal soil stress can be obtained with the pressiometer and/or dilatometer and the deformation characteristics can be determined more or less directly. The execution of pressiometer tests is rather time consuming. Dilatometers are pushed down in the same manner as the CPT. This type of investigation usually only makes sense when the stability of adjacent structures is at stake.

Density investigations
The best information on the packing density is obtained with the investigation as described in Chapter 11. However this type of investigation is seldom required for information on the borrow area.

Twin well probe
The local permeability of the soil can be measured with the twin well probe. From the outside the shape of the probe is identical to the lowest part of a CPT and is pushed in the soil in the same manner. In the soil directly around the probe a ground water flow is initiated by pumping water in the ground at one point and by sucking up in the probe the same volume of water at a distance of 0.5 to 1.0 m from that point. At various locations along the probe the hydrostatic pressure is measured. From this the permeability of the surrounding soil can be derived.

Seismic investigation
Shallow seismic investigation based on reflection and/or refraction is suitable to rapidly identify the vertical position of a layer separation along a line. The separation between rock and sediments can almost always be established. The separation between different types of sediments, for example clay and sand, requires the results of some borings along the line of the seismic survey for correlation. Even than the layer separations cannot always be established everywhere due to lateral variation of sediments.

3.3   Extent and phasing of investigations

The extent of a soil investigation in the potential borrow areas will generally be limited, particularly when compared with an investigation for the foundations of buildings, quay walls, caissons and the like. This limitation is related in the first place to the type of information required as discussed in the beginning of this Chapter. The strength and the strain parameters are only required to be known with restricted accuracy.

For this reason the necessity to carry out many measurements per unit area is also restricted: variations in these parameters are of less importance than for investigations involving foundations. Often only the variation of the soil structure, the layer thickness and some essential soil characteristics such as grain size distribution and the density of the soil are of relevance. How many measurements per unit area (one per 10 hectare or 10 per hectare) are required in connection with this variation, can only be established after one has already obtained some knowledge of the area.

For reason given above it often makes sense to carry out some initial geological research and subsequently carry out a limited number (say 10) of bore holes and classify the retrieved samples. After that the matrix can be refined with seismic survey, more borings with SPT's and/or cone penetration tests. The extent of detailing must be based on the expected variation of the layer thicknesses and the essential soil characteristics. Geostatistical methods can be of use for quantifying these expectations. Finally supplementary investigations may be carried out if necessary for the benefit of specific soil characteristics such as density or undrained shear strength.

CHAPTER 4

FALLING OF A SAND-WATER MIXTURE THROUGH WATER

4.1 Hydraulic sand placing processes

4.1.1 *Methods of hydraulic placing*

In practice a variety of techniques of hydraulic placement of sand is employed. The most commonly used techniques are single point discharging in an underwater fill area, whereby a sand-water mixture is placed in the works via a delivery pipe, and the discharging of sand from a hopper barge. If single point discharging of sand in an underwater fill area is adopted, then the outlet of the delivery pipe may either be positioned just above the (sea) bottom or above water level. If the discharging takes place from a barge, this is either via valves or doors or by opening a split dumping barge. When discharging via valves or doors in the bottom of a hopper (barge/dredge), the sand flows down like a sand-water mixture in the shape of a curtain with a width equal to the width of one valve and a length equal to the total length of the row of valves in the vessel. In the case of a split dumping barge, the opening in the vessel is very large causing the sand to fall in large lumps.

These methods will respectively be referred to as curtain-like discharging and lump-like discharging (see also 2.3). However it is doubted whether lump-like discharging really occurs with a split dumping barge. Presumably the opening of the split dumping barge usually takes enough time to allow the lump to form a hyperconcentrated thin wisp or curtain so that the process more or less will resemble curtain-like discharging.

The following parameters are of influence on the process of hydraulic sand placement:

- discharge height;
- density of the sand-water mixture;
- cross–section of the discharge opening;
- mixture flow rate;
- sand characteristics.

It is mainly during the stage of sand placement that designers and contractors can influence the consecutive processes. Once the method of sand placement and accompanying circumstances have been selected most of the boundary conditions of the process are defined.

4.1.2 *Fall processes*

The fall processes can be classified as follows (see Fig. 32):

- the jet/plume mechanism, which describes the processes that occur during single point discharging under water and during curtain-like discharging;

47

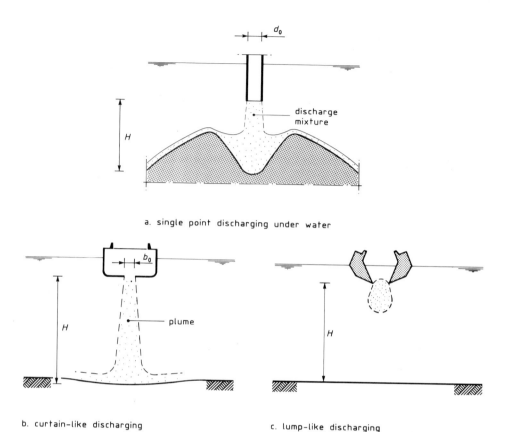

a. single point discharging under water

b. curtain-like discharging

c. lump-like discharging

Fig. 32. Types of fall processes.

– the mechanism of a falling sphere, which may be used to describe the dumping process when employing a split dumping barge.

The jets and plumes are the result of the fact that the discharged mixture has a (considerably) higher velocity than the surrounding fluid. This difference in velocity causes turbulence which allows the discharged mixture to entrain water from the surrounding fluid. This entrainment increases the diameter of the jet, resulting in its typical V shape. However, although the shape of a jet or plume may be similar, the mechanisms differ fundamentally: the jet flows due to the initial momentum of the mixture, whereas for the plume the difference in density with the surrounding fluid is the driving force. This implies that differences in density are not required to create a jet; an example of this is a pure water jet in water. An example of a plume is the smoke of a cigarette. A combination of mechanisms is also possible: a so called jet driven plume. In this case both types of driving forces are relevant: nearby the discharge opening the jet mechanism prevails whereas at a greater distance the plume mechanism governs the mixture behaviour.

48

Each of the above described mechanisms may occur in the following two flow shapes: the conical shaped jet and the plane vertical shaped jet. The conical or round shaped jet develops when a flow leaves a circular opening (thus for single point discharging in an underwater fill area), whereas the vertical plane shaped jet develops for a rectangular opening (hence for curtain-like dumping).

The most commonly encountered falling processes are either the conical jet driven plumes in case of single point discharging under water or the vertical plane shaped jets in case a dumping barge is used.

When a split dumping barge is used for the discharging of sand under water, this can most probably be described as curtain-like discharging, but it could also be that the sand falls down in several large lumps (lump-like discharging). These lumps may schematically be represented as spheres (see Fig. 33). If such a sand sphere is dropped in the water, its mass changes during the fall because sand whirls off the sphere in trailing vortices and simultaneously water is entrained into the sphere. During the fall these mechanisms changes the diameter, shape and density of the sphere.

The sand losses of the sphere are determined by the falling velocity, the diameter and a constant which can be calibrated by measurements. The water entrainment is the result of internal vortices, caused by the difference in velocity between the sphere and the surrounding water [2].

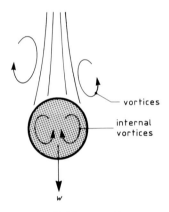

Fig. 33.   Falling sand lumps.

4.1.3   *Impact processes*

The sand-water mixture reaches the area of impact, often a sea or river bottom, with a certain density, size and velocity. In this paragraph the processes during impact are described in a qualitative manner for the different methods of discharging in water. In Chapter 5 these are discussed in a quantitative manner.

Single point discharging under water

When observing the impact area during single point discharging under water it appears that the jet digs out a so called whirl crater in which the sand-water mixture flows turbulently. The momentum of the mixture continuously erodes the center of the crater. However on the slopes the velocity reduces allowing the sand to settle. As sand continues to settle, the crater gradually moves to a higher level whilst the volume remains more or less constant.

The size of the crater and the time required to create a fully developed crater depends on the equilibrium between erosion and deposition. A few experimental observations are available, but the processes themselves are only known in broad outlines. Reference is made to Chapter 5 for the calculation of the dimensions of a whirl crater.

Curtain-like discharging
During curtain-like discharging no whirl crater has been observed in which the mixture flows around. In the area of impact often a wide shallow crater develops mainly because the current hits the bottom over a very large area. The vertical plane shaped jet is deviated and flows after impact as a horizontal suspension flow. Therefore the velocity, flow profile and density of the deposition-flow next to the area of impact can be derived directly from the theory of vertical plane shaped plumes.

Whether a berm develops or a crater depends on the equilibrium between erosion and sedimentation.

Quantitive information is only available on clear water jets and plumes (see Chapter 5).

Lump-like discharging
The shape of the area of impact for discharging sand depends very much on the boundary conditions: how big are the lumps initially, what are their densities and in what water depth are they (bottom) discharged.

Because the density of the sand sphere decreases during the falling process it is possible that at greater depth the falling lump of sand transforms into a round plume flow. Therefore the shape of the area of impact may range between a berm and a shallow crater. At this moment for both the falling processes and the impact processes there is only information available on small lumps of approximately ten centimeters in diameter. For quantitative conclusions more research has to be carried out on the behaviour of prototype-size lumps.

4.2   Jet/plume mechanism

Jet trajectory
If the fall velocity W_0 of the individual sand grains is small compared to the velocity U_0 of the jet, the sand-water mixture can be considered as a jet with a density which differs from the ambient fluid (buoyant jet). A practical value is: $U_0/W_0 > 50$. For a large fall velocity of the grains the sand-water mixture cannot be considered as a jet. The initial momentum and the difference in density are the driving forces of the buoyant jet.

50

The jet trajectory can be divided into two sections:

- The zone of flow establishment (ZFE) where the transition takes place from the conditions during discharging to a Gaussian velocity and concentration distribution. The length of the ZFE is about 6 to 10 times the nozzle diameter.

- The zone of established flow (ZEF). In this zone it is assumed that the cross sectional distribution of velocity and concentration remains Gaussian.

At a „large" distance from the discharge point the velocity of the jet in relation to the surroundings and the turbulence has been reduced to such a level that the behavior of the surrounding fluid will dominate the flow. The assumption of a buoyant jet is then no longer valid.

Empirical formulae
When the distance is greater than 5 to 10 times the diameter of the nozzle, empirical formulae can be used to determine the flow velocity and density of the jet [2, 3]. The trajectory of a jet in a flowing medium cannot be determined with these formulae.

Integration method
The velocity, density and trajectory of the jet in the zone of established flow, can be calculated by solving the formulae which describe the process. The description can be simplified by integrating the variables over the cross-section of the jet. The relative distribution of the velocity and density is assumed to be Gaussian. Conservation of mass and momentum is also assumed.
The interaction between the jet and the surrounding fluid is schematically represented as a drag force, a buoyancy force and the entrainment of the surrounding fluid. A density and velocity distribution of the surrounding water can be taken into account (see [4] for more details on the method of calculation).
To illustrate this method an example is given:

A sand-water mixture is vertically discharged in flowing water under the following circumstances:

- diameter of nozzle: $d_0 = 0.8$ m
- discharge velocity: $U_0 = 4$ m/s
- velocity of surrounding water: $U_a = 1$ m/s
- mixture density during discharge: $\varrho_0 = 1300$ kg/m^3
- density of surrounding water: $\varrho_a = 1025$ kg/m^3
- vertical distance, bottom to nozzle: $H = 18$ m

The equations are solved numerically with the computer program „STRAAL3D" developed by Delft Hydraulics. In Figure 34, the velocity and density in the center line of the jet are shown as a function of the vertical distance z to the discharge point as well as the horizontal distance x from the center line of the jet to the discharge point. The volume concentration in the center line of the jet follows from:

$$c = \frac{\varrho_m - \varrho_w}{\varrho_s - \varrho_w}$$

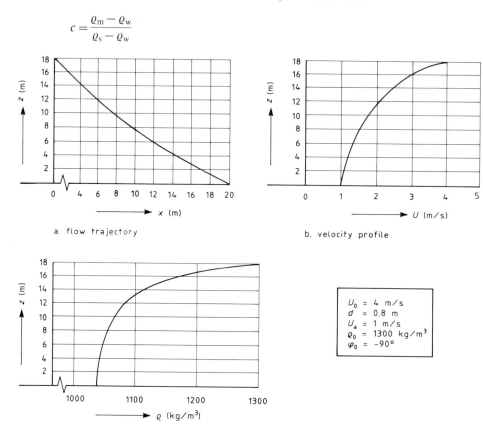

a. flow trajectory

b. velocity profile

c. density profile

$U_0 = 4$ m/s
$d = 0.8$ m
$U_a = 1$ m/s
$\varrho_0 = 1300$ kg/m^3
$\varphi_0 = -90°$

Fig. 34. Calculation results of the computer program „STRAAL3D".

The jet diameter can be defined as:

$$d = d_0 \sqrt{\frac{c_0 U_0}{c U}}$$

In which:

d = jet diameter at height z above the bottom (m)
U = flow velocity of the jet at height z above the bottom (m/s)

52

CHAPTER 5

FORMATION OF CRATERS

5.1 Dimensions of a whirl crater

As a result of the impact of the sand-water mixture plume or jet an erosion crater develops. In this Chapter some formulae are given to estimate the dimensions of the crater. There are two approaches. The first begins with the description of the erosion under a clear water jet followed by the determination of the influence of sand in the jet on the dimensions of the crater. With a second, alternative approach the crater dimensions are directly determined from tests and experience with sand-water jets. Finally an example is given. In 5.2 some remarks are made with respect to the time scale. In 5.3 an estimate follows on the spreading of the mixture that flows across the crater edge.

5.1.1 *Erosion by clear water jets*

Conical or round jets
The approach of BREUSERS [5,6] is used to determine the ultimate crater dimensions (see Fig. 35).
An analysis has been made of the trajectory of a round jet impinging on a flat plate (see Fig. 36). In this case the following relation [7] applies between the velocity U_m on the plate and the velocity U_0 being the discharge velocity from the delivery pipe:

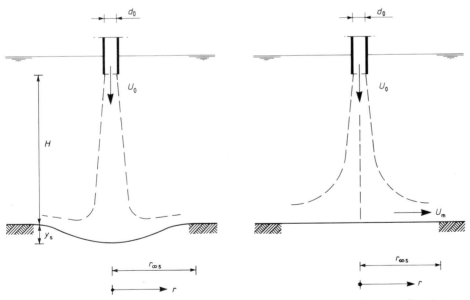

Fig. 35. Erosion under a vertical jet.   Fig. 36. Jet impinging on a flat plate.

$$\frac{U_m}{U_0} = 1.03 \frac{d_0}{r} \tag{1}$$

The velocity along the surface of the plate decreases with distance r. For small values of r the velocity is large enough to erode material, whereas for increasing r sedimentation occurs. A simplified approach for the determination of the dimensions of the scour hole may be obtained by assuming that at a distance $r_{\cos}$ from the center line of the jet the flow velocity U_m equals a certain value U_c. When the flow velocity is greater than U_c, erosion occurs. The relation between U_c and the critical shear velocity u_{*cr} is reflected in:

$$u_{*cr}^2 = \frac{f_0}{8} U_c^2 \tag{2}$$

The radius of the crater can be found by combining the formulae (1) and (2), where $U_m = U_c$:

$$\frac{r_{\cos}}{d_0} = 1.03 \frac{U_0}{U_m} = 1.03 \sqrt{\frac{f_0}{8}} \cdot \frac{U_0}{u_{*cr}} \tag{3}$$

It appears further that the ratio between the radius $r_{\cos}$ and the depth y_s of a scour hole is constant under a vertical mixture jet and about equal to 2.5. In formula (3), f_0 is a Darcy-Weisbach friction coefficient and is of the order of 0.02 to 0.1. In this way it is found that:

$$\frac{r_{\cos}}{d_0} = 0.05 \text{ to } 0.12 \frac{U_0}{u_{*cr}}$$

$$\frac{y_s}{d_0} = 0.02 \text{ to } 0.05 \frac{U_0}{u_{*cr}}$$

Based on experiments (see Fig. 37), BREUSERS [5] recommends the following slightly deviating relations:

$$\frac{y_s}{d_0} = 0.075 \frac{U_0}{u_{*cr}} \qquad \text{if} \quad \frac{U_0}{u_{*cr}} < 100$$

$$\frac{y_s}{d_0} = 0.35 \left(\frac{U_0}{u_{*cr}}\right)^{2/3} \qquad \text{if} \quad \frac{U_0}{u_{*cr}} > 100 \tag{4}$$

In which:

$\quad\quad d_0 \ = \text{the discharge pipe diameter}$
$\quad\quad U_0 \ = \text{the discharge velocity}$
$\quad\quad u_{*cr} = \text{the critical shear velocity according to SHIELDS (see Fig. 38)}$

For sand u_{*cr} is 0.013 m/s ($D_{50} = 100$ μm) to 0.025 m/s ($D_{50} = 2000$ μm).

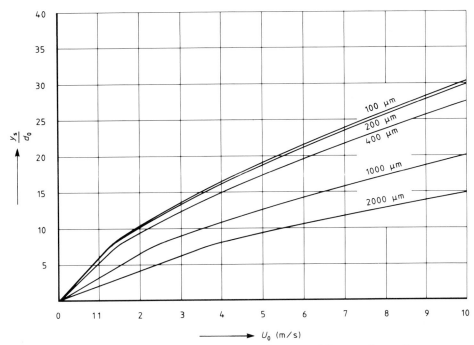

Fig. 37.   Depth of scour hole for clear water as function of U_0 according to BREUSERS.

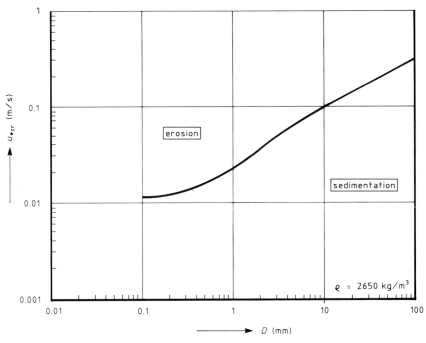

Fig. 38.   Critical shear velocity as a function of the grain size diameter.

The vertical distance H to the point of discharge has no influence, at least within the application range of the formula. This range is:

$$\text{approximately } 4 < \frac{H}{d_0} < 12 \text{ to } 20$$

For smaller values of H/d_0 the crater dimensions become smaller.

Vertical plane shaped jets (for curtain-like discharging)
RAJARATNAM [8] provides the following formula for the erosion under a vertical plane shaped jet based on experimental data:

$$\frac{y_s}{2b_0} = 0.23 \frac{U_0}{\sqrt{g \, \Delta D_{50}}} = \sqrt{\frac{H}{2b_0}} \tag{5}$$

Reference should be made to Figure 39 for clarification of this formula. The experiments were carried out with a clear water jet on a sand body with a D_{50} of 1200 μm and 2400 μm, velocities U_0 up to approximately 3 m/s and values of $H/2b_0$ from 5 to approximately 50 (N.B. $2b_0$ was approximately 1.5 mm).

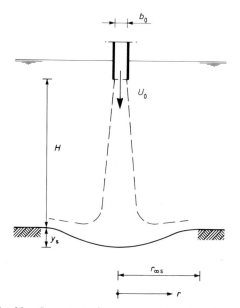

Fig. 39. Scour hole for vertical plane shaped jets.

Formula (5) is shown in a graphical form for the situation $H/2b_0 = 50$ in Figure 40 for different grain diameters. Also the experimental results found by RAJARATNAM are shown for the two types of sand mentioned and for the various values of $H/2b_0$ applied during the tests. Hence for higher values of U_0 and lower values of D_{50} extensive extrapolation has to be carried out.

56

The formulae of RAJARATNAM and BREUSERS are based on small scale laboratory experiments. These formulae have not been verified for a large scale. However, based on the knowledge of hydraulic processes and subsequent experience (scour holes) extrapolation to a larger scale seems to be justified. It is obvious that the results are limited.

5.1.2 Effect of sand in a water jet on the dimensions of a crater

As described in 5.1.1, the dimensions of a scour hole are determined by an equilibrium between erosion and sedimentation. In the case of a clear water jet these can be found with the formulae (4) and (5).

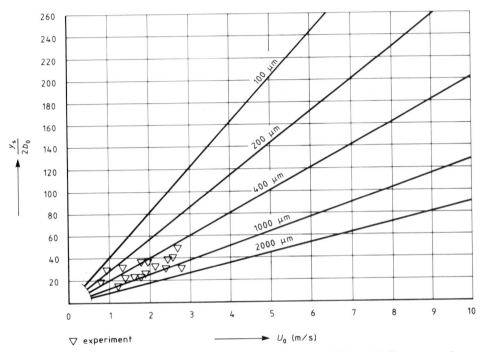

Fig. 40.   Crater depth under a vertical plane shaped jet for $H/2b_0 = 50$ (RAJARATNAM).

The effect of sand in the jet can be found again by analyzing the trajectory of a jet impinging on a flat plate, assuming equilibrium between erosion and sedimentation. Because of the presence of sand in the flow, the velocity of the flow on the crater edge must be higher than for clear water in order to prevent sedimentation. As a result of this the crater edge will be closer to the center line of the jet.

The presence of sand can be accounted for, by reducing the crater dimensions, calculated with the formulae (4) and (5), using a reduction factor which depends on the grain size diameter D_{50} and on the sand concentration c in the jet. In Figure 41 this reduction factor has been plotted versus the grain size diameter and the concentration. For

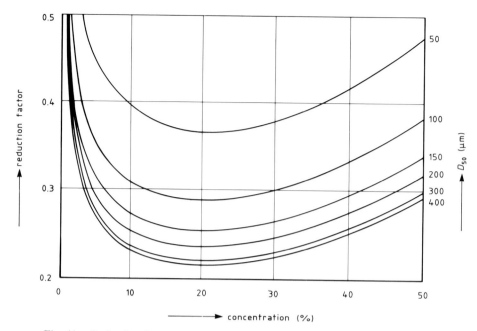

Fig. 41. Reduction factor as a function of concentration and grain size diameter.

example, it appears from this Figure that for jets with coarse sand ($D_{50} = 400$ µm) a reduction of the dimensions of the crater up to a factor 5 is possible.

The ratio between radius and depth does not change, hence the following always applies:

$$\frac{r_\infty}{y} = \frac{r_{\infty s}}{y_s} = 2.5 \tag{6}$$

5.1.3 Erosion by round sand-water jets; alternative approach

Another method to determine the geometry of a crater under a sand-water jet is based on the results of small scale experiments. As described in [2], it was found that for discharge tests close to the sand bottom ($1 < H/d < 4$), HEEZEN and VAN DER STAP found that the ratio of the momentum of the jet and the mass of water inside the crater is more or less constant. This ratio is described by the dimensionless parameter HS. Using formula (2) this formula can be written in the following form:

$$\frac{y}{d_0} = 0.4 \left\{ \frac{1}{HS} \cdot \frac{\pi(1 + \Delta c_0)U_0^2}{4gd_0} \right\}^{1/3} \tag{7}$$

In which:

$$\Delta = \frac{\varrho_s - \varrho_w}{\varrho_w}$$

ϱ_s = density of the grains

58

The value of *HS* for whirl craters, developed during hydraulic discharging, is approximately 0.015 to 0.03 for sand with a grain size diameter of 150 to 320 μm. In this formula the reduction caused by the presence of sand has been accounted for.

The indicated values apply to the developed scour hole after the jet or plume has stopped. During single point discharging the scour depth can be greater and the inner slope can be steeper. However the scour diameter does not changed.

Note: If the discharge opening is kept at a great distance above the bottom $(H/d_0 > 4)$ the equation can also be applied provided U_0, c_0 and d_0 are replaced by U, c and d as determined with the jet calculations (see 4.2.1) at bottom height. This formula is shown graphically in Figure 42.

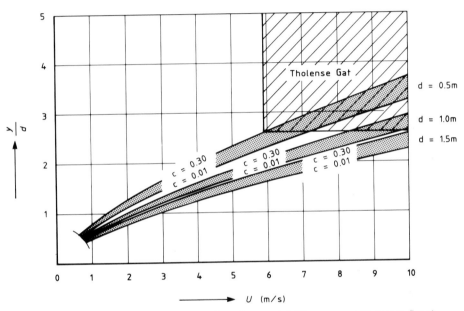

Fig. 42.   Crater depth under a round sand-water jet (HEEZEN and VAN DER STAP).

5.1.4   *Example*

During the construction of the Oesterdam across the Tholensche Gat in the province of Zeeland, The Netherlands, sand was discharged into the underwater fill area through a delivery pipe which was vertically suspended from a movable spray pontoon. Here, craters have been measured with a diameter of 10 to 25 m. The shifting of the discharge opening along the submerged berm was possibly an important cause for the large variation in diameter. More detailed data are not known to the authors [23].

These prototype measurements may be compared with calculations using the different theories described in this Chapter.

The following situation applies to the Tholensche Gat case:

- grain size: $D_{50} = 185$ μm
- diameter of the discharge opening: $d_0 = 0.9$ m and 1.0 m
- mixture flow rate: $Q_0 = 4$–6 m³/s
- mixture flow velocity during discharge: $U_0 = 6$–7.5 m/s
- volume concentration: $c_0 = 0.21$
- mixture density: $\varrho_m = 1350$ kg/m³
- density of the surrounding water: $\varrho_w = 1000$ kg/m³
- sand production: 2200–3000 kg/s

Calculations provide the following result:

Using the „HS" theory, with a value of the HS parameter of 0.03, the calculated crater diameter is approximately 12 m; this is within the measured range (however somewhat on the low side) of 10 to 25 m.

Using formula (4) or Figure 37 with the reduction factor from Figure 41, a calculated diameter follows of 32 m which is somewhat on the high side but of the same order of magnitude as the measured value.

5.2 Time scale of erosion processes

In the preceding paragraphs the final crater dimensions are presented which develop after some time after reaching an equilibrium situation. The final crater dimensions are often based on experimental model investigation.

The result of those models can be translated to prototype dimensions using the following general relation between the erosion depth and time as derived from small scale experiments [9] on local erosion:

$$\frac{y}{L} = f\left(\frac{t}{t_1}\right) \qquad (8)$$

In which:

L = characteristic length (for instance jet diameter)
y = erosion depth
t_1 = time required for reaching the erosion depth for which $y = L$

When L_p and L_m are the respective lengths in the prototype and in the model, U_p and U_m the respective velocities in the prototype and in the model then the ratio t_p/t_m between the prototype time and the model time for reaching a certain erosion can be found using the relation:

$$\frac{t_p}{t_m} = \left(\frac{L_p}{L_m}\right)^2 \left(\frac{U_m}{U_p}\right)^{4.3} \qquad (9)$$

With this relation the required time can be predicted for reaching a certain erosion depth on prototype scale, when results of model tests or small scale works are available. This relation can also be used to scale up the experience of designers and contractors. Without model tests the upper limit of the crater depth which develops when single point discharging during a period Δt or when bottom discharging can be estimated by assuming the following:

- The crater width immediately reaches its ultimate size $2r_{\infty}$;
- The maximum volume of sand which erodes from the crater during a certain period Δt equals the volume of the sand-water mixture entering the crater during the same period. The maximum crater volume would then be equal to $U_b \frac{1}{4}\pi d_b^2 \Delta t$. The real crater volume will probably be much less.

5.3 Mixture overflow from crater

After a crater has been formed in the original sea bottom or in the already placed sand under the influence of the mixture flow from a discharge outlet, the mixture flow will pass the crater edge across a certain width. The question is, across what width B the mixture will spread. There is little known about this subject. For that reason the following approach must be regarded as a first attempt to estimate an upper limit B_{max} and a lower limit B_{min}. The upper limit is taken equal to the circumference of the crater edge. In principle B could become even larger after the mixture has passed the crater edge because the size of the crater increases continuously from the discharge point. However it seems very unrealistic that the mixture spreads more than the complete circumference of the crater edge.

Hence, for a crater under a round (conical) jet the following applies:

$$B_{max} = 2\pi r_{\infty} \tag{10a}$$

and for a crater under a vertical plane shaped jet this with a length equal to the curtain-length or the hopper-length of the vessel from which the sand is discharged, the following applies:

$$B_{max} = 2\pi r_{\infty} + 2 \times \text{curtain-length} \tag{10b}$$

However most likely the mixture does not spread across the complete crater circumference B_{max}, but will somewhat concentrate in a channel. On the fill area above water it has been observed that the mixture flow tends to form a channel with rather high flow velocities and specific flow rates. This channel shifts continually because of sedimentation during the run off (see Fig. 43).

On the abovewater fill area the mixture flow can be distributed by means of bulldozers across the full width of the fill area determined by bunds. This is not the case on the underwater fill area and hence channel formation cannot be prevented here.

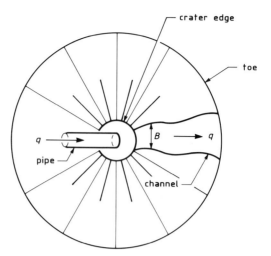

Fig. 43. Mixture overflow from crater.

The mixture flow now behaves as a density current for which the driving force is determined by the difference in density with the surrounding water and thus by the concentration. It is assumed that the density current flows across the crater edge and forms a channel analogous to the situation on the abovewater fill area. For vertical pipeline placement, this channel shifts continuously so that ultimately the mixture flow covers the complete circumference of 360°. The width B of this channel determines the specific flow rate q and the specific sand production rate s and therefore the underwater slopes and the possibility of flow slides.

In order to calculate the minimum channel width B_{min}, it is assumed that the mixture flow behaves in the same way as on the abovewater fill area and does not spread any further after the crater edge is passed. It is known from the regime theory [9] that the width B of a channel is proportional to the square root of the total flow rate Q; this regime theory was established from observations of relatively small irrigation channels. From measurements on the fill area of the Speelmansplaten (Marrollegat, Delta works) [15] it is further known that the channel width varies from 3.8 to 5 m for a discharge rate of 2.9 m³/s and a grain size diameter of 150 μm. From this the width of a channel can be derived, normalized for jet diameter d (for a round jet, hence for single point discharging in an underwater fill area):

$$\frac{B_{min}}{d} = 0.6 \sqrt{\frac{U}{U_c}} \qquad (11)$$

In which:

B_{min} = channel width (m)
d = jet diameter (m)
U = mixture flow velocity in centre of jet (m/s)
U_c = mixture flow velocity for which erosion occurs, (SHIELDS:
      $U_c \equiv \sqrt{\Delta g D_{50}}$ (m/s)

In Figure 44 this relation is plotted as a function of the flow velocity in the centre of the jet and the grain size diameter.

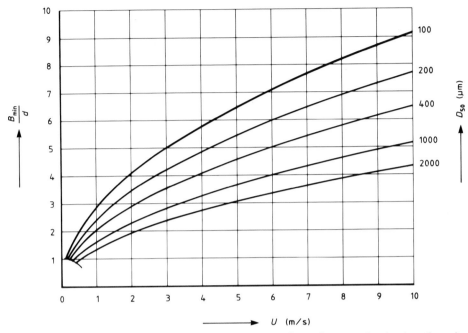

Fig. 44. Minimum spreading width of mixture at overflow of crater edge (regime theory).

An alternative approach, which is more specifically aimed at the situation on the under-water fill area, assumes that the flow across the edge of the scour hole is critical, meaning that it behaves as a spillway with internal Froude number equal to 1. From research on breach formation in a sand body during a dike-breach [10] it appeared that the adapted width/depth ratio for critical flow is always equal to 4 approximately. From this the following can be derived for single point discharging in an underwater fill area:

$$\frac{B_{min}}{d} = 2 \frac{U^{0.4}}{d^{0.2}} \left(\frac{1 + \Delta c}{\Delta c g} \right)^{0.2} \tag{12}$$

Figure 45 shows this relation graphically. The occurrence of critical flow is now also determined by the mixture density, hence the concentration is also of importance. Besides that it appears that B_{min} is not exactly proportional to the jet diameter as in the previous formulation.

The results of both formulations overlap each other to a large extent, both relations show a proportionality with the discharge velocity to the power 0.4 to 0.5. As the mixture flows further along the crater slope it may be expected that the internal Froude number of the flow in a channel decreases gradually, during which the flow velocity will decrease at more or less constant mixture depth. It appears then that the width/depth

63

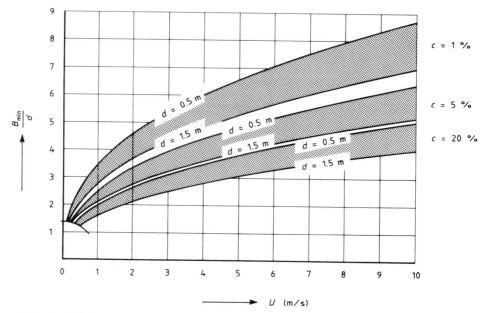

Fig. 45. Minimum spreading width of mixture at overflow of crater edge (critical flow).

ratio increases up to a value of approximately 10 to 15, as has been observed in the channels of the Marollegat.

When the discharge opening is just above the crater $(H/d_0 < 4)$ the formulae (11) and (12) and the Figures 44 and 45 can be used with: $U = U_0$, $d = d_0$ and $c = c_0$. In other cases $(H/d_0 > 4)$ values of respectively U_b, d_b and c_b on bottom level (height of crater edge) have to be filled in, which follows from the method of calculation of 4.2. The specific flow ratio q and specific sand production rate s for single point discharging in an underwater fill area, follow now from the formulae:

$$q = \frac{U_b \frac{1}{4} \pi d_b^2}{B} \tag{13}$$

$$s = \varrho_s q c_b \tag{14}$$

For curtain-like discharging the formulae (11) and (12) and the Figures 44 and 45 can be used provided d has been replaced by:

$$d = \sqrt{\frac{4b \cdot \text{curtain-length}}{\pi}}$$

CHAPTER 6

SEDIMENTATION AND EROSION PROCESSES

The extension of a hydraulically placed sand body is the result of sedimentation of sand from a sand-water mixture flow and in some cases the successive replacement of already deposited sand.

6.1 Equilibrium flow

With the Figures given in this Chapter, both sedimentation and erosion processes can be determined in a quantitative way. In principle these processes are physically independent of each other. Hence the net result can therefore be derived by super-imposing the sedimentation and erosion velocity. Under special circumstances a net equilibrium can exist between erosion and sedimentation. In that case the net erosion or sedimentation velocity equals zero and the height of the sand body will not change in the coarse of time.

The most important parameter which determines the equilibrium situation is the average slope of the sand body (see also the Chapters 7 and 8). Generally it can be stated that for flat slopes the sedimentation processes prevail and for steep slopes the erosion processes.

During the hydraulic sandfill process three types of sand-water mixture flows can be distinguished, namely grain flow and suspension flow, which are of a hydraulic nature, and flow slides which are of a soilmechanical nature.

6.2 Sedimentation during grain flow

Grain flow is the laminar flow of thin hyperconcentrated sand layers with a thickness of a few millimeters, on steep slopes of 1:2 to 1:3. Grain flows occur for small specific sand production rates with values less than 5 kg/ms. The grains in the flowing layers are dispersed by mutual grain to grain interactions. Sedimentation of grains from this type of flow occurs through regular deposition of thin layers of sand grains within a short distance. This process of sedimentation looks like the sudden „freezing" of the flowing sand tongues [12, 13].

6.3 Sedimentation during suspension flow

Suspension flows behave like a turbulent density flow in which the sand is completely kept in suspension by turbulent vortices. Suspension flow occurs for moderate to high specific sand production rates with values beyond 5 kg/ms [13, 14].

When a turbulent sand-water mixture flows over a loose sand bed, a continuous inter-action takes place between the flow and the sand bed. For this type of flow some typical hydraulic bedding formations on the fill area can be distinguished such as terrace formation and hydraulic jumps.

Sand may deposit from the flow, but simultaneously sand may also erode from the bottom if the local flow velocity is sufficiently high.

During hydraulic placing of sand, the sedimentation generally will be greater than the erosion, resulting in a net sedimentation. As a result the sand body will expand.

Sedimentation from a suspension flow per unit of time and per unit of area, or the so called vertical downward directed flux per unit of area, is defined as the product of the fall velocity of the sand grains in the suspension flow and the volume concentration. The fall velocity of single grains is a function of the grain size and, not to be overlooked, of the water viscosity and therefore also of the water temperature [15].

The fall velocity in water of single grains of quartz sand (the most common type) with a density of 2650 kg/m³ as a function of grain size and temperature, can be read from Figure 46. The influence of minor differences in water-density on the fall velocity, for instance salty water versus clear water, can be neglected.

The effective fall velocity of sand grains in hyperconcentrated mixture flows, as encountered during hydraulic sand fill processes, is strongly reduced in comparison with the fall velocity of a single grain. This is because neighbouring grains hinder the

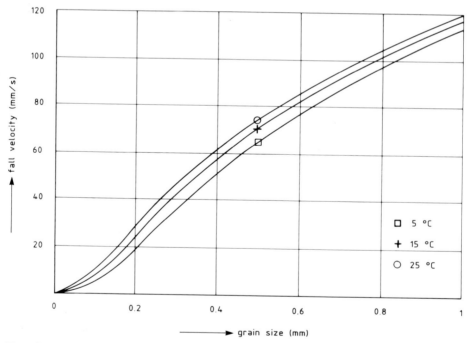

Fig. 46. Fall velocity of sand grains in still water as a function of grain size and temperature.

sand particles to settle. Hence this process is called „hindered settling" and is a function of sand concentration. This effect has clearly been demonstrated in the laboratory by the registration of concentration profiles in hyperconcentrated sand-water mixture flows [16].

The sedimentation from a sand-water mixture flow can be expressed in terms of sedimentation rate being the velocity at which the height of the sand bed increases. In Figure 47, the sedimentation rate is given as a function of grain size and (initial) concentration, for a temperature of 15 °C and a porosity of the sand bed of 44 %. Because of the effect of hindered settling the sedimentation rate shows an optimum which is reached at a volume concentration of approximately 20 %. The latter corresponds with a mixture density of approximately 1350 kg/m^3.

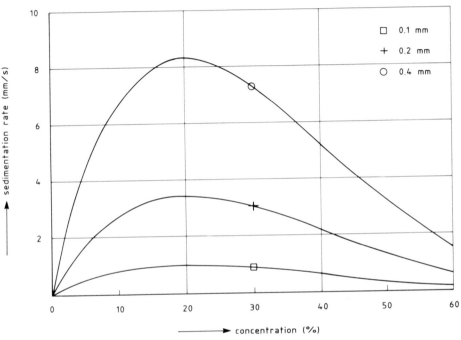

Fig. 47. Sedimentation rate from a sand-water mixture flow as a function of grain size and volume concentration.

During the sedimentation of the sand-water mixture flow, sand with pore water is deposited as a result of which the flow, the sand concentration and the resulting sand transport decrease.

The sedimentation length is defined as the distance covered by the sand-water mixture flow before the volume concentration is reduced to 10 % of the initial concentration. This sedimentation length can be read from Figure 48 as a function of the specific mixture flow rate, the grain size and the volume concentration. Because of the hindered settling effect, the sedimentation length is strongly influenced by the concentration.

Since the sand production rate is the product of mixture flow rate and concentration, the sedimentation length is strongly determined by this parameter. In practice, the specific sand production rate together with the grain size diameter are the most important parameters of the process of hydraulic filling.

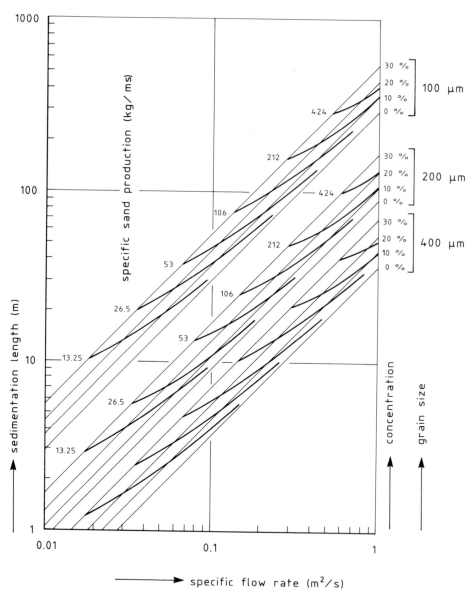

Fig. 48. Sedimentation length as a function of specific flow rate, grain size and volume concentration.

Note 1.

The sedimentation length according to Figure 48 is based on the presence of a continuous mixture flow. If this flow is only present during a very short period, than the sedimentation length will be shorter.

Note 2.

The influence of the grain size in Figure 48 is based on the assumption that it concerns homogeneous sand (one grain size diameter). For sand with a flat sieve curve, the small grains have a relatively great influence. Hence, a value must be chosen for the typical grain size which is smaller than D_{50}, say for instance D_{15}.

6.4 Sedimentation during flow slides

The sedimentation length of grain flows and suspension flows at small and moderate specific sand production rates, is generally small in relation to the length of the slope of the sand body. This results in a locally increasing slope gradient with a high porosity in which flow slides may occur [13, 14]. The liquefied mass of sand, with a porosity almost equal to the porosity of the sand body (40 to 50 %), moves along the slope by means of laminar flow, as a very viscous fluid resulting in an enormous sand transport in a very short time (within some minutes). This type of flow of a weakened sand body is called a flow slide. The liquified sand grains will settle again analogously to the way indicated in Figure 47. After the flow slide a sand bed is obtained with a slightly lower porosity and a flatter slope. The process of alternating increase and decrease of the slope, caused respectively by sedimentation of a grain or suspension flow and by flow slides, goes on continuously during the hydraulic sand fill process. Therefore for short sedimentation lengths, in comparison with the dimensions of the sand body, flow slides can be the dominating process for the sand transport to the toe and consequently decisive for the final slope. This has been observed during hydraulic sand filling with $D_{50} = 120$ to 250 μm. However at large sedimentation lengths, the suspension flows determine the process (see further 8.4 and 9.4).

6.5 Erosion

Although the sedimentation process is the most important phenomenon during sand fill operations, erosion of the sand bed may occur in some situations. For instance the occurrence of the formation of craters at the end of the delivery pipeline, the erosion of bunds and the erosion downstream of a terrace shaped fill area combined with a slow upstream propagation of these terraces [16].

The erosion process in hyperconcentrated sand-water mixtures is of a completely different nature compared to similar processes in rivers or canals.

It has been proven that there is a maximum value of the specific sand transport rate at a given mixture flow rate, because the volume concentration cannot raise beyond the value defined by the minimum porosity. For this reason the increase of the concentration and the specific sand production rate is limited at increasing bottom resistance or mixture velocity. The exchange of pore water and sand between the sand bed and mixture, necessary for the erosion process, will be hindered by the already high values of the concentration. This effect is called „hindered erosion" which is analogous to the hindered settling effect on the fall velocity of grains in a mixture.

Another important effect is caused by the local slope gradient of the eroded sand bed. At steep slopes the erosion increases even for a constant bottom shear stress. When the local slope is equal to or even exceeds the angle of internal friction (angle of repose), the erosion process is determined by soil mechanics processes and not by the mixture velocity. Figure 49 shows the erosion velocity, being the opposite of the sedimentation velocity, as a function of the mixture velocity, the local slope gradient (with a maximum of approximately 32°, being the angle of repose of the sand in question) and, for the maximum erosion velocities, the concentration.

Figures 46 to 49 are calculated with the computer program ZSTORT, which is based on both mathematical and physical formulations supplemented with empirical relations found during detailed investigations on hyperconcentrated sand-water mixture flows [15].

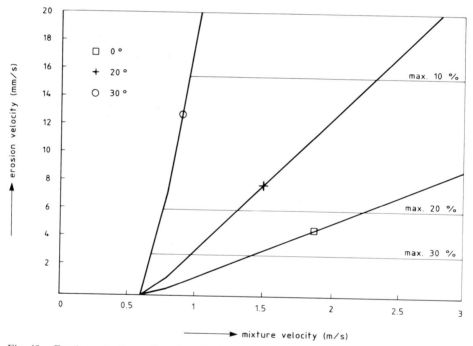

Fig. 49. Erosion velocity as a function of mixture velocity, volume concentration and local slope gradient for $f_0 \approx 0.05$ and 125 μm $< D <$ 200 μm.

CHAPTER 7

HYPERCONCENTRATED SAND-WATER MIXTURE FLOW ABOVE WATER

7.1 Methods of hydraulic placement

The construction of a hydraulically placed sand body can be achieved by two essentially different placement methods, namely:

a. placement in a horizontal direction (= pipeline placement above water level);
b. placement in a vertical direction (= pipeline placement below water level or curtain-like discharging or lump-like discharging).

These methods of hydraulic placement are described in Chapter 2. In case a the sand-water mixture flow, which is hydraulically supplied via a pipeline from a hopper suction trailing dredger or suction dredger, flows beneath and above the water level. The sand body develops in a horizontal direction by sedimentation of sand from the mixture flow (see Chapter 6). This is also the case for the construction of a sand fill on land.

Hydraulic discharging of sand for the closure of the Krammer.

When discharging in a vertical direction, as mentioned under b, the sand body will primarily develop in a vertical direction.

Above water level the sand-water mixture behaves more or less like an ordinary water flow, whereas under water the flow will behave as a density current.

7.2 Types of sand-water mixture flow

A sand-water mixture flow consists of a fluid phase and a solid phase. The type of flow is defined by the characteristics of the fluid, the solids and the flow.
Typical hydraulic properties, concerning the fluid phase are for instance the viscosity and the density.
Typical properties related to the solids are for instance the concentration, the grain size and the density.
Typical properties related to the flow are for instance the flow velocity and the turbulence or energy dissipation.
All cases concern mixtures of which the volume of solids is certainly not negligible in comparison with the volume of water. Hence the term „hyperconcentrated" in the title. In this Chapter two types of flow are distinguished namely laminar or grain flow and turbulent or suspension flow (see Chapter 6). In practice both these types of sand-water mixture flow can be present. Above water however the flow velocities usually are sufficiently high for turbulent flow.

Grain flow

Under laminar flow conditions no vertical transport takes place (by definition) between the layers of the mixture flow. Grain flow occurs at low specific mixture flow rates (less than about 0.01 m^2/s) in thin layers with a thickness of only a few sand grain diameters, at relatively high flow velocities. The sand grains in the grain flow are supported by interactions between the grains. It is also possible that the viscous fluid keeps the grains in suspension [13]. During small scale flume experiments, performed at Delft University of Technology, grain flow phenomena could be clearly observed [17].

Suspension flow

In a turbulent mixture or suspension flow an intensive diffusive exchange of water and sand grains takes place in a vertical direction. Turbulent flow occurs at moderate and high specific flow rates (more than about 0.05 m^2/s). The turbulent nature of the flow will be influenced by the presence of the grains. Through interactions between the grains and the fluid, turbulent excess hydrostatic pressures are developed.
The vertical concentration distribution and the net sedimentation rate in the suspension flow (see Chapter 6), are determined by the fall velocity of the grains in the mixture and the degree of turbulence. Owing to the strongly reduced fall velocity (hindered settling), very large quantities of sand grains can be kept in suspension at relatively moderate turbulence levels, which explains the observed extremely high specific sand transport rates.
Large scale flume tests were performed at Delft Hydraulics [16]. Some important properties of the hyperconcentrated turbulent suspension flow could be established such as concentration and velocity distributions [16, 18].

72

Velocity distribution

The velocity distributions tended to be fairly logarithmic (see Fig. 50). At high concentrations, the thickness of the viscous sub-layer near the bottom seemed to increase, which resulted in an increased apparent friction coefficient. This phenomenon can probably be explained by the increased fluid viscosity (internal friction) caused by the presence of the grains.

Another remarkable effect was the shift and rotation of the logarithmic profiles at varying concentration. This is induced by an initial decrease and subsequent increase in the damping of the turbulent movements with increasing sand concentration. A maximum effect occurs at $c \approx 20$ %.

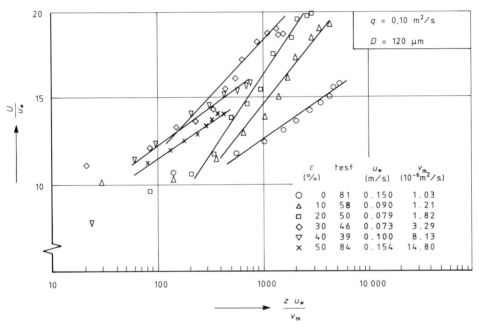

Fig. 50. Vertical velocity distributions of a mixture flow.

Concentration distribution

The concentration profiles for fine sand (120 μm) (see Fig. 51), showed well mixed almost homogeneous profiles, especially at concentrations beyond approximately 20 %. For medium coarse sand (225 μm), a stronger gradient was observed and higher flow velocities appeared to be required (through a steeper slope) to obtain uniform flow. This phenomenon is caused by the greater fall velocity of the grains in the suspension flow.

7.3 Bed-form types on the hydraulic sand fill

In practice [17,19] the following bed-forms can be observed: terraces, cascades and channel formation.

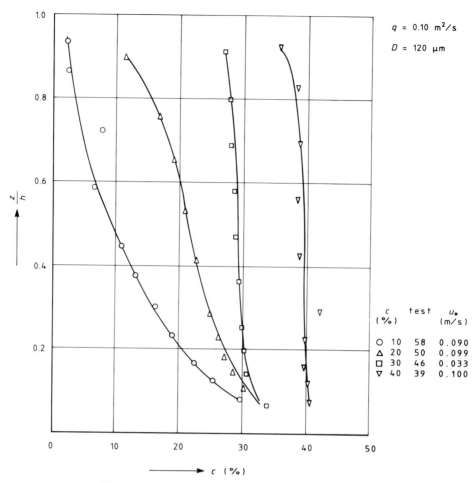

Fig. 51. Concentration distributions in a mixture flow.

Terraces and cascades

Terraces and cascades develop at moderate specific mixture flow rates of the order of 0.01 to 0.3 m²/s. The anti-dune like bed-form is composed of long, flat sloped terraces where the flow velocity is sub-critical and where sedimentation of sand takes place (see Fig. 52). At the end of this terrace a steep cascade is present, where the flow velocity becomes supercritical and where erosion takes place. A hydraulic jump [19] develops between this cascade and the next terrace at the transition of supercritical flow to sub-critical flow. Because of the erosion of the downstream side, the system of terraces, steps and cascades propagates slowly in upstream direction with a velocity of approximately 0.01 to 0.03 m/s. A dynamic equilibrium situation can develop between sedi-

74

mentation on the terraces and erosion on the cascades at a certain average bed slope. This bed-form type develops over the full width of the slope.

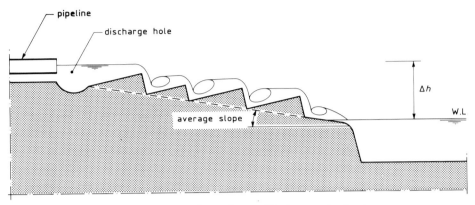

Fig. 52.   Terrace formation on fill above water level.

Channels
At high specific mixture flow rates the mixture concentrates in deep channels and the bed-form of terraces and cascades disappears. A new equilibrium situation has developed with more or less uniform flow at flatter slopes.
The developed bed-form depends on the average Froude number, on the average bed slope and on the sand characteristics (Shields number, [14]).

7.4   Equilibrium slope

The equilibrium slope is defined by the equilibrium between sedimentation and erosion (see Chapter 6). The equilibrium slope could be determined with large scale tilting flume tests [16] as a function of the flow velocity, the grain size and the concentration. An empirical formula [18,20] for this equilibrium slope above water level could be inferred. In simplified form:

$$i = 0.006 \left(\frac{D_{50}}{D_0} - 1 \right) \frac{q^{-0.45}}{q_0} \tag{15}$$

In which:

i = tangent of the average slope gradient
D_{50} = medium grain size diameter (μm)
q = specific mixture flow rate (m²/s)
D_0 = 65 μm
q_0 = 1 m²/s

75

Terrace formation on abovewater fill area.

Channel formation on abovewater fill area.

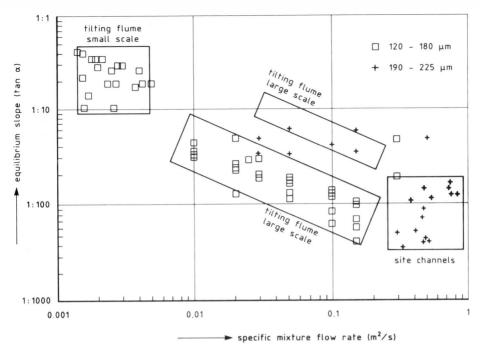

a. results of flume tests and field observations

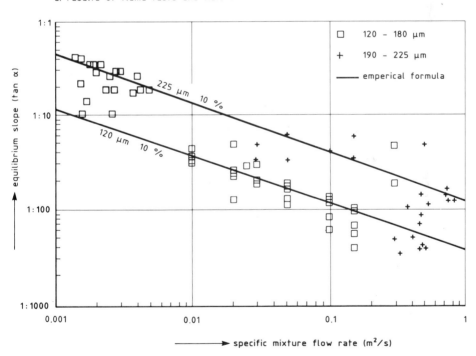

b. results of flume tests and field observations
and calculations using formula (15)

Fig. 53. Equilibrium slopes.

This formula was derived for specific mixture flow rates varying from 0.01 m²/s and 0.15 m²/s and grain size diameters of respectively 120 μm and 225 μm.

In Figure 53b these relations are shown together with the flume and field surveys (see hereafter). The equilibrium slope showed a strong relation with the specific mixture flow rate and the grain size. The concentration turned out to be of minor importance, which can be explained by the effects of hindered settling (see Chapter 6).

This effect is also present in the mixture flows in the pipeline or in the fill area, where the critical velocity does not depend on concentrations if these are higher than about 15 %, where the influence of hindered settling comes into effect.

Another comparison can be made with sand transport formulae for channel or river flow, for instance the formula of ENGELUND-HANSEN. By accounting for the effect of hindered settling this formula can be modified. It shows great resemblance with the empirical formula (15) [20].

The anti-dune type of bed-form has been developed in the flume at a somewhat flatter slope than the equilibrium slope. Also during net sedimentation on the fill area the overall slopes appeared not to be much different from the equilibrium slopes.

The equilibrium slopes for laminar grain flow above water, determined with small scale flume tests were, in contrast to the foregoing, only dependant on the concentration according to the formula:

$$i = \frac{c}{1 + 1.65c} \tag{16}$$

In which:

$$c = \text{volume concentration of the sand}$$

This behaviour can be understood by considering the grain stress distribution in a uniform flow [18].

Field surveys of bed slopes were performed during mixture flow in channels [17] as well as during mixture flow on terrace-cascade bed-form types [19]. Figure 53a gives the measured equilibrium slopes as a function of the specific mixture flow rate during small scale and large scale flume tests and during the field surveys. In Figure 53b the values are also shown as lines calculated with the unsimplified version of the empirical formula (15), as a function of the specific mixture flow rate, the grain size and the concentration. A reasonable correlation with the field observations is found.

With the computer program ZSTORT for hyperconcentrated sand-water mixture flows [15], not only the dimensions of the bed forms and hence the dimensions of the whole fill area can be determined, but also the dynamic equilibrium slope.

CHAPTER 8

HYPERCONCENTRATED SAND-WATER MIXTURE FLOW
UNDER WATER

8.1 Methods of hydraulic placement

The construction of a hydraulic fill area in water can be realized by placement in a horizontal direction or by placement in a vertical direction. For placement in a vertical direction distinction can be made between pipeline placement in a subaqueous fill area, curtain-like discharging and lump-like discharging. This has been described briefly in 7.1 and more detailed in 2.1 and 4.1. For each of the placement methods hyperconcentrated sand-water mixture flows occur under water.

8.2 Types of sand-water mixture flow

The development of a sand body under water is the result of sedimentation processes. Sand settles from the various types of sand-water mixture flows. Under water the same types of mixture flows are present as above water, namely grain flow and suspension flow (see Chapters 6 and 7). Here again the attention is restricted to hyperconcentrated mixtures although underwater suspension flows with low concentrations may occur in the vicinity of the falling and flowing mixtures, sometimes visible as turbid clouds. Compared to the circumstances above water however, the soil mechanics properties of the sand body during sand placement under water have a much stronger influence on the development of the slope. Because the deposited sand is always very loosely packed, it is susceptible to liquefaction. Therefore under certain circumstances the slope of the sand body is not stable and liquefaction may occur followed by a flow slide.

The behaviour of mixture flow under water is mainly determined by the specific sand transport rate instead of the mixture flow rate above water. The average value of the vertical distribution as well as the concentration is of importance.

Whether turbulent or laminar flow occurs depends on the Reynolds number. At low specific sand transport rates ($s < 10$ kg/sm) and for fine sand, grain and suspension flows of small thickness occur. For fine sand as well as for coarse sand however, a turbulent suspension flow occurs at high specific sand transport rates ($s > 25$ kg/sm).

8.3 Bed-form types on the underwater hydraulic sand fill

Similar to the bed-form types above water a terrace-cascade system with hydraulic jumps will develop propagating continuously in the upstream direction. In some cases the cascade height increases due to strong erosion. A rather high breach will then develop and move upstream. The eroded sand settles again beyond the internal hydraulic jump. Eventually this breaching process results in a flat downstream sand slope comparable to the result of a flow slide.

The length over which the suspension flow can extend under water is approximately equal to the sedimentation length (see Chapter 6).

When all the sand has settled and no sand is being eroded, the driving force of the density current disappears and the flow declines.

8.4 Slope development under water of a sand fill

To illustrate the development of a sand slope under water as described here, reference is made to Figure 54. The parameter L mentioned in this Figure is defined at the end of this paragraph. For the development of the slope the size of the specific sand production rate is of importance as well as the grain diameter (see also 6.4 and 9.4).

Low to moderate sand production rates
During hydraulic placement of sand at low to moderate specific sand production rates and with fine sand in shallow to fairly deep water, the development of the slope takes place in a discontinuous way. This has been observed during field surveys [20,21] and during experiments in a large flume [13, 22].

During the hydraulic placement of sand, sedimentation takes place from a grain or mixture flow within the first metres of the slope after the sand body has reached the water surface. It then starts to develop in horizontal direction. The slope increases locally which may result in a critical situation and a flow slide (see Fig. 54a).

Sedimentation again takes place on the resulting flat slope. The alternating increase and decrease of the slope caused by sedimentation and flow slides, is a continuous process. The flow slides are now the dominating factor for the transportation of sand to the toe of the sand body. More about this type of slope formation can be found in 9.4. For coarser sand and for the same low specific sand production rates and shallow water, however, flow slides are observed less readily even when the sand is very loose and the porosity of the deposited sand is well beyond the critical value.

It was found that for horizontal slope delopment a critical water depth exists. Beyond this critical height flow slides occur. For coarse sand this critical slope height h_{cr} is larger than for fine sand (see Fig. 55). The values for h_{cr} are based on experience with the construction of sand fills with grain sizes between 100 µm and 500 µm. The critical slope height can be approximated by:

$$h_{cr} = 0.075D - 8.5 \tag{17}$$

In which:

h_{cr} = critical slope height (m)
D = grain diameter (µm)

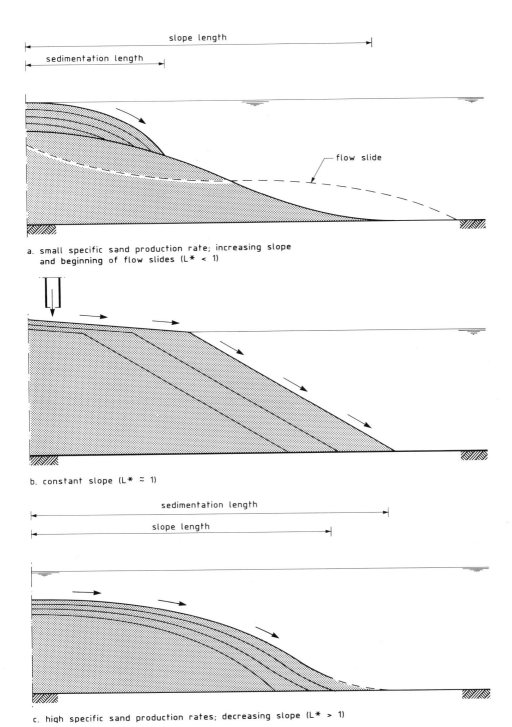

a. small specific sand production rate; increasing slope
   and beginning of flow slides (L* < 1)

b. constant slope (L* ≈ 1)

c. high specific sand production rates; decreasing slope (L* > 1)

Fig. 54.   Development of sand slopes.

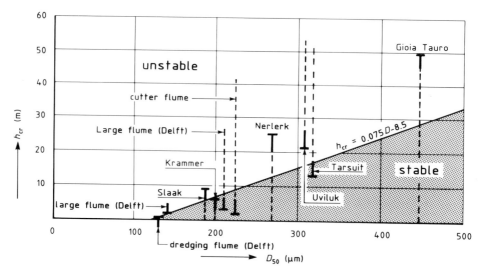

Fig. 55. Critical slope height.

This formula gives merely a rough indication. Also the slope and the porosity at initial sedimentation are of influence. As this slope is flatter and its porosity is smaller, h_{cr} will be larger. Compare also the criteria for flow slides in natural deposited sand as described in 9.3. There most of the time $h_{cr} \geq 5$ m, whereas a critical sand slope is assumed.

For $h < h_{cr}$ the underwater sand slope increases up to a certain steep equilibrium angle. The development of the slope takes place in a very regular way and a kind of equilibrium situation establishes (see Fig. 54b).

High specific sand production rates

For high specific sand production rates ($s > 25$ kg/ms) a turbulent suspension flow occurs, for fine sand as well as for coarse sand, with hydraulic jumps propagating continuously in the upstream direction. This process is very similar to the type of mixture flow above water. The suspension flow can extend far beyond the toe of the slope resulting in a decreasing underwater slope (see Fig. 54c). Although flow slides may be present these do not constitute a prevalent sand transport mechanism.

Dimensionless parameters

In order to quantify the three described types of slope development three dimensionless parameters can be distinguished.

In the first place a dimensionless hydraulic sedimentation parameter L^* can be defined being the ratio between the sedimentation length of the sand-water mixture and the length of the sand slope beneath water level (see Fig. 54):

$$L^* = \frac{\text{sedimentation length}}{\text{length of the sand slope}} \tag{18}$$

82

If this ratio is small, sand will settle on the upper reach of the slope and the slope gradient increases. Especially if fine sand is present a flow slide then occurs (see Fig. 54a).

If L^* is large, sand flows beyond the toe of the slope and the gradient of the slope decreases (see Fig. 54c).

For in between values, $L^* \approx 1$, the slope will remain more or less constant (see Fig. 54b). Also two dimensionless soil parameters can be defined, which indicate whether liquefaction will or will not occur.

The first parameter N^* is defined as the ratio between the actual and the critical porosity:

$$N^* = \frac{\text{actual porosity}}{\text{critical porosity}} \tag{19}$$

Also for the height of the sand slope a soil parameter can be defined. This second parameter H^* is defined as the ratio between the actual and the critical slope height (formula (17)):

$$H^* = \frac{\text{actual slope height}}{\text{critical height}} \tag{20}$$

As observed during the field surveys and the flume tests the porosity of the settled sand will always be well above the critical value during the process of hydraulic placement of sand. Therefore almost always $N^* > 1$ is valid and hence the sand body always is susceptible to liquefaction.

During small scale and large scale flume tests [13,22] it has been observed that with the same fine sand and the same high porosity liquefaction did occur in a large deep flume ($H^* > 1$) but not in a shallow flume ($H^* < 1$). Moreover, in the same large flume and with the same sand production rate liquefaction did occur with fine sand and not with the coarser sand. Apparently the critical slope height increases with the grain size (see Fig. 55).

8.5 Sand fill slopes beneath water level

As described in 8.4, no unique value can be established for the slope of a hydraulically placed sand body because of the alternating increase and decrease of the slope. Flow slides or breach processes, if present, define the minimum slopes. The average underwater sand slope is mainly defined by the grain size and the specific sand production rate.

Average slopes observed during the hydraulic placement of sand together with the regression lines for these average slopes, are shown in Figure 56 as a function of the specific production rate, the grain size and the construction method. The results of flume tests with fine and medium coarse sand are given for various sand production rates, as well as the results of previously performed small scale flume tests [14, 20, 22]. The results of field observations of various hydraulic sand fill projects [20, 23, 38], are included in the Figure as well.

Figure 56a shows the slope gradient as a function of the specific sand production rate and Figure 56b shows this slope gradient for the same data as a function of grain size. Since all types of measurements are collected a wide scatter is the result. From the Figures 56a and 56b it appears clearly that the specific sand production rate is of less influence than the grain size.

From large scale flume tests [14,22], which give a more distinct relation, it could be concluded that the specific sand production rate and grain size were important parameters. The concentration or the specific flow rate appeared to be of minor importance. The following empirical formula has been inferred for the average underwater slope:

$$i = 0.0032 D s^{-0.4}$$

(21)

In which:

i = slope gradient of sand body
D = grain size (μm)
s = specific sand production rate (kg/ms)

The sand production rate was varied between 2 kg/ms and 80 kg/ms, the grain size between 135 μm and 225 μm, the specific flow rate between 0.005 m^2/s and 0.1 m^2/s, the concentration between 12 % and 30 % and the water depth as well as the slope height was about 2 m. Formula (21) is shown in the Figures 56a and 56b for different values of D and s.

As mentioned in 7.4, the equilibrium slope of a sand-water mixture under water can be inferred analogously using the well known formula of ENGELUND-HANSEN [13]. Again a formula very similar to formula (21) is obtained. Under water however, the specific sand production rate s is the governing parameter whereas above water this is the specific flow rate q (see formula (15)).

It can be read from the Figures that in the field often flatter slopes are observed than according to formula (21). These are always practical situations for which flow slides govern the slope formation. For these practical situations, for which always $L^* < 1$ and $N^* > 1$ and $H^* > 1$ (see 8.4), it is therefore recommended not to use formula (21) but to estimate the slope on the basis of the indicated field observations.

8.6 Sand losses during hydraulic sand placement

Another aspect is the exchange of sand and water between the sand-water mixture flow under water and the ambient fluid. The mixture flow can dilute and sand can be lost to the surrounding water by entrainment and diffusion. If a water flow occurs near the sand slope perpendicular to the density current, the entrained sand may be transported outside the limits of the sand body under construction and must be considered as a loss.

84

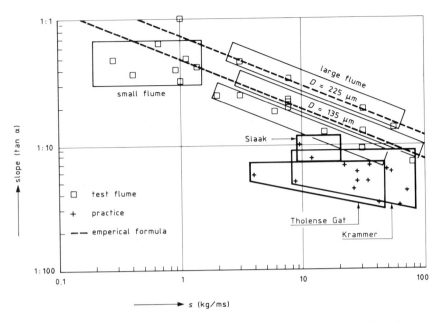

a. under water sand slopes as a function of the specific sand production rate

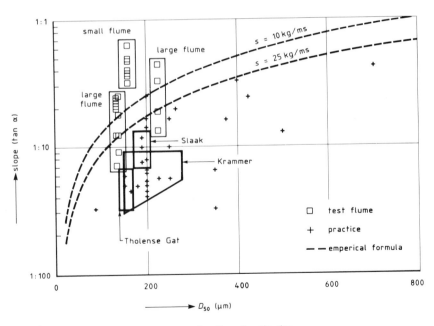

b. under water sand slope as a function of grain size

85

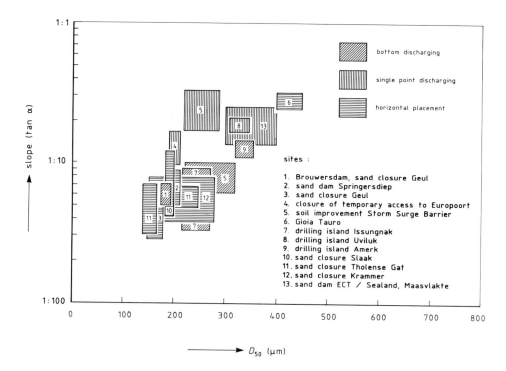

c. under water sand slopes as measured in the field as a function
of grain size and construction method

Fig. 56. Underwater sand slopes as measured in the fild and in the laboratory.

The volume of these sand losses depend on the entrainment rate, determined by the flow velocity and the area of the slope covered by the mixture flow. As shown before, the sand usually settles on the top side of the slope and will be transported downstream by flow slides. In that case the area is only limited and the sand losses due to the mixture flow are small.

However, the flow across the sand slope may erode again already deposited sand. This erosion may increase considerably at increasing flow velocities [24]. These effects can be determined with the usual sand transport formulae.

The contribution of the entrainment of sand from the sand-water mixture flow to the total sand loss is generally only small.

CHAPTER 9

FLOW SLIDES

9.1 General

The material in this Chapter is mainly derived from LINDENBERG [25] and SILVIS [26, 27].

The phenomenon of a flow slide is the liquefaction of a volume of sand in an underwater sand slope and the subsequent flowing of the sand along the slope to a lower level. Here, the following sub-processes can be distinguished, which can take place after each other but also at the same time:

– liquefaction of sand;
– mixture formation;
– mixture flow;
– sedimentation of sand.

In Figure 57 the sub-processes and their location to each other along the slope are indicated.

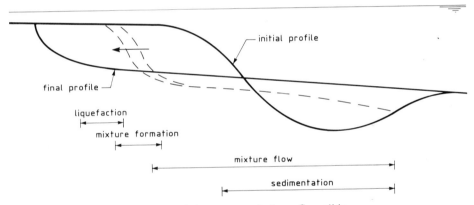

Fig. 57.   Sub-processes during a flow slide.

The volume of sand which flows away can sometimes be very large. In the province of Zeeland in the Netherlands flow slides involving 2,000,000 m^3 of sand have been observed.

Two other types of instabilities of the underwater slope do exist, namely slip failures and breach formation. During the latter process a steep part of the slope (breach) moves upward along the slope. This may also be called caving flow.

These types of instabilities can only develop at steep slopes. After instability the slope usually is much steeper than the one developed after a flow slide. For these type of instabilities excess pore pressures do not play a role or only a minor one. Breaching can only occur in dense sand, for example during cutter suction dredging. Instead of pore overpressures, pore underpressures occur. Slip failures and breach formation only occur in the sand borrow pit and usually do not constitute a problem. For that reason only the phenomenon of flow slides is discussed in this Chapter.

9.2   Explanation of the flow slide phenomenon

Flow slides can only develop in loose sand saturated with water.
Loose sand has the tendency to decrease in volume at increasing shear stress (see Fig. 58a) and denser packing of the grain structure will result. Dense sand however has exactly the opposite tendency: the packing becomes looser, this is called dilatation (see Fig. 58b).

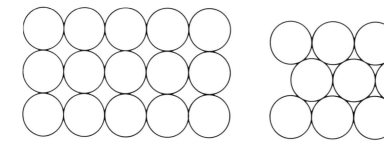

a. loose condition   b. dense condition

Fig. 58.   Loose and dense condition of sand.

If the sand is saturated with water and practically no water can flow in or out, the said change of volume cannot take place. Instead overpressure or underpressure respectively occurs. As a result, the effective stress decreases respectively increases. Under certain circumstances for loose sand, the effective stress can be negated completely by the excess pore pressure. The grains float in the water so to say: this is a matter of a liquified sand-water mixture which behaves as a heavy viscous fluid. This liquefaction only occurs if the water cannot flow away fast enough. This is why flow slides mainly occur in relatively fine sand.
From the foregoing it follows that increase of shear stress is required to cause liquefaction. Increase of shear stress and subsequent deformation is initiated by a triggering mechanism such as a change of the sand slope or a vibration.
The triggering mechanism can be caused by the suction in a borrow pit and by cutter dredging of the subaqueous slopes. Examples of this are the problems of slope dredging during the harbour extensions in the Sloe (Vlissingen, the Netherlands) and in the Eems/Dollard (Delfzijl, the Netherlands). Furthermore the triggering mechanism can be caused by the seeping out of ground water, by sand fills (for example during the

extension of a sand body), shockloading and vibrations as a result of pile driving, explosions, earth quakes or direct wave impact.

However in most of the Dutch cases, presumably most of the time, it is a matter of deformation of the grain structure due to the increased steepness of an embankment caused by the erosion of the lower reach of the slope, the deepening of a channel or the sedimentation of the upper reach of the slope. Little is known of experiences outside the Netherlands.

9.3   Practical advise for naturally settled sand

At present it is generally accepted in practice, that for the development of a flow slide three of the following conditions must be satisfied:

1. the sand must be sensitive to liquefaction across a minimum layer thickness which varies presumably between 5 m and 10 m;
2. the sand slope must be steeper than a certain critical value over a sufficient height (often also 5 m to 10 m); compare also h_{cr} for hydraulically placed sand, as given in 8.4 (formula (17));
3. There must be some kind of a triggering mechanism: for instance change of slope, water level and or soil stress or vibration.

sub 1. Sensitivity to liquefaction
At present it is possible to check whether the first condition is satisfied. For this, first of all geological information must be collected. This may provide an indication on sensitivity to liquefaction. In this way it appears that in the Netherlands flow slides almost only occur in Holocene sand. However, seldom can firm conclusions be drawn and therefore (disturbed) samples are recovered of the relevant soil. The (wet) critical density of reconstituted samples can be determined in a laboratory. At the same time the density in the field is measured or assessed through penetration tests. If these densities are smaller than the critical density, the sand is sensitive to liquefaction.

sub 2. Exceedance of the critical steepness
Recently a theoretical method has been developed to check whether the second condition is satisfied. For this, the deformation characteristics of the sand have to be determined first by means of laboratory tests on (disturbed) samples. A calculation model is then examined to determine whether sudden excess hydrostatic pressures can occur for the given geometry. In order to examine this, the initial stress situation of the sand must be known in principle. It's true that this is difficult to measure but nevertheless this can be estimated for instance by means of penetration tests. At present the accuracy of this estimate is still limited. Therefore, at present, the theoretical method is only suitable to establish a safe lower limit to the steepness of the slope.

Except by means of this theoretical method, using their experience, experts can also establish whether the second condition is satisfied. The reliability of establishing the critical steepness is greatest for areas as Zeeland in the Netherlands, where extensive well documented experience is available [28, 29, 30, 31]. As a result the critical slope steepness appears to be between 1:3 and 1:7.

sub 3. Triggering mechanism
The third condition is almost always satisfied if it concerns an unprotected slope: a triggering mechanism is always likely to occur although the moment itself is difficult to predict.

These conditions only concern the question whether flow slides will develop and what section of the sand slope is involved. The question as to where the sand will flow has yet to be answered. For that knowledge is required about the sub-processes of mixture formation, mixture flow and sedimentation (see Fig. 57). Last mentioned sub process can be described quantitatively (see Chapter 6). This does however not apply to the other two sub-processes. The prediction of the eventual profile after a flow slide, is therefore a matter of experience. As an average over the final profile, slopes result of 1:5 to 1:15.

From experience some knowledge exists on the influence of „storage" (places where sand can flow to) on the final profile.

9.4 Flow slides during hydraulic placing of fill under water

This subject is also dealt with in 6.4 and 8.4. During the (horizontal) hydraulic placement of a sand dam the required sand-water mixture is supplied via a pipeline across the already completed section of the dam (single point discharging above water). The discharged sand settles above water level as well as below water level. From measurements at the Marollegat (Delta works, the Netherlands), it appeared that nearly all of the sand that ends up under water, settles initially on the upper reach of the underwater slope. Nevertheless, a great deal of the sand ends up at the lower reach of the slope. This can be clarified by the development of several flow slides.

In order to verify this assumption an extensive monitoring campaign was launched during the hydraulic placement of the sand dam in the Slaak in 1986. This is one of the compartment dams in the Eastern Scheldt Estuary (the Netherlands). During the hydraulic sand placing process various types of measurements were performed from a platform in the channel, such as: hydrostatic pressures, porosities, sand concentrations and movements of the sand surface. Everything indicated that sand transport takes place towards the toe from a level directly below the water line, through a variety of many rather shallow flow slides with depth of approximately 1 m [25]. At a later stage the same thing was observed during large scale model tests (see 6.4 and [13]).

The observed process can briefly be described as follows. The sand which arrives at the water line settles (for the first time) at the upper part of the underwater slope in a very loose state, with a porosity n being sometimes greater than the maximum value found in the laboratory: $n > n_{max}$. If the specific sand production rate – the sand quantity reaching the water line per unit of width and time – is small, the sand settles directly below the water line. A very steep slope develops at the upper part of the underwater slope (see for example the sedimentation between I and II in Fig. 59).

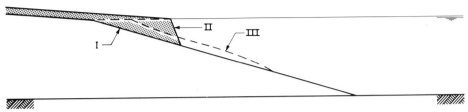

Fig. 59.   Development of a slope profile of a sand fill under the influence of flow slides.

If the specific sand production rate is larger, then it will be a matter of a longer sedimentation length. This sedimentation length is the length over which the sand in suspension would settle, if there were no turbulent diffusion. Because of this greater length it's true that the sand would settle at a greater distance from the water line, but would still end up at the upper part of the underwater slope. Very steep slopes will still develop locally, with rapid changing soil stresses.

As soon as the sand level is raised through sedimentation with the critical height h_{cr}, as defined in Chapter 8, the pore pressure suddenly increases strongly in that area. (At the compartment dams h_{cr} amounted to approximately 1 m). Across a height of at most approximately h_{cr}, the sand volume is completely liquefied or at any rate sufficiently liquefied in order to loose stability and to start to flow (see Fig. 60). The excess pore pressure increases roughly in a linear way, from the surface to the bottom side of the liquefied layer. Beneath that, the excess hydrostatic pressure gradually decreases.

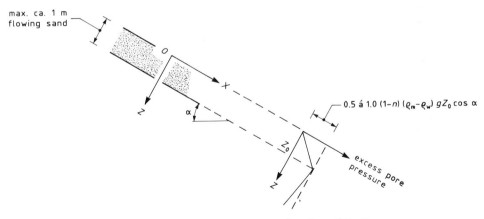

Fig. 60.   Excess pore pressure as a function of depth.

The said mixture of sand and (ground)water slowly starts to move. Almost at the same time, the lower sand layers start to settle (for the second time) as described in Chapter 6. With that, the groundwater is squeezed out and the sand achieves a slightly denser packing. It can last up to one minute before all the sand is settled again. The required period of time is directly proportional to the thickness of the liquefied layer. Therefore the trend in the time of the porosity and the excess pore pressure looks for instance as indicated in Figure 61, in which the trend of the pore pressure and the decrease of the porosity is given for a fixed point situated about 0.5 m below the surface of the slope. In this Figure it can be observed that the process repeats itself a number of times. The resulting slope profile is marked with III in Figure 59.

Presumably such flow slides always occur in rather fine sand ($D_{50} < 200$ μm) and seldom in coarse sand ($D_{50} > 300$ μm to 400 μm) (see also formula (17) and Fig. 55). This could explain why slopes for dams built with coarse sand are so much steeper than dams built with fine sand and agree better with the slope according to formula (21) (see Fig. 56b and [20]). Due to flow slides, the sand density is never as low as directly after the first sedimentation. From measurements on hydraulically placed sand fills it appears however, that the density after the occurrence of a flow slide is often just beyond the critical density.

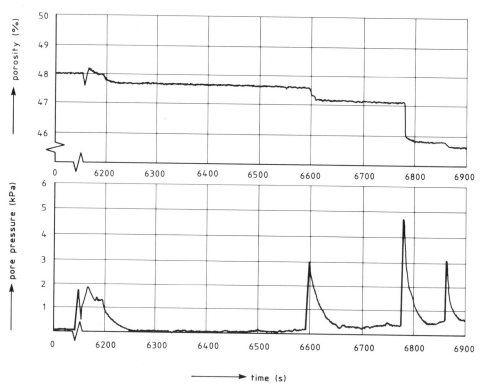

Fig. 61.   Example of increase of the pore pressure and decrease of the porosity during a flow slide.

Therefore it is not impossible that the sand is sensitive to liquefaction, after the completion of the sand fill. However, usually no more flow slides will occur after construction owing to the usually flat sand slopes.

9.5   Measures to prevent flow slides

9.5.1   *Compaction*

The sensitivity to liquefaction of sand is, to a great extent, connected with the density or the porosity of the sand. By compaction of the soil to a porosity lower than the critical porosity, the sensitivity to liquefaction is removed. After that, the further occurrence of flow slides is not possible. Soil compaction was extensively applied in the trajectory of the Eastern Scheldt storm surge barrier (deep compaction by means of the vibrating-needle vessel Mytilus) where the result was very satisfactory.

A problem with the deep compaction through vibration of in principle liquefiable sand is, that this goes together with created excess hydrostatic pressures in the soil and eventually even liquefaction.

This means that deep compaction in a location sensitive to flow slides cannot be applied directly along the (critical) slope, because it is very likely that a flow slide will be initiated by the compaction activities.

9.5.2   *Slope regrading*

The occurrence of a flow slide can also be prevented by adding material to a slope (section) up to provide a gradient smaller than the critical slope, often combined with the careful placing of gravel or rock along the complete slope from the lower end to the upper end. For this purpose, often the placement of coarse material is applied (for instance: embankment of Zuid-Tholen along the Tholense Gat, Delta works). At the same time such a protection prevents further erosion and steepening of the slope. This is often the main purpose rather than slope grading as such (condition 2). The occurrence of a sliding or other triggering mechanisme is prevented as well (condition 3).

9.5.3   *Measures during the hydraulic placement of sand*

It is unfortunately not clear as to whether and how a greater density can be achieved during the pipeline placement or discharge process. Probably the process itself can be of influence. If for instance one would operate with small sand flow rates and would distribute these flow rates across the fill area in such a way that flow slides will be prevented during the hydraulic placing and steep slopes would develop, then there would be a risk of a very loose packing and hence a big chance on flow slides at a later stage.

Laboratory tests [2] also give an indication of the influence of the placement method.

The authors arrive at the conclusion that a higher density will be achieved when the grains settle one by one and are given the chance to search for the deepest niche. This is not possible when large groups of grains settle at the same time. This principle has also been successfully applied during the closure of the temporary mouth of the Europort. There, the sand was brought to the bottom via a vertical pipe, suspended from a pontoon which was continuously shifted. The thin layers placed in this way, were for a long time subjected to a weak tidal current so that the sand grains only settled when finding a niche.

From a thorough comparison between the various artificial islands, SLADEN and HEWITT [33] observe that the relative density of bottom dumped sand is considerably larger than that of sand supplied by pipeline placement. The same appeared from penetration tests and density measurements during the construction of the Brouwersdam in the Netherlands. Possibly it's true to say that the relative density is low after first sedimentation, but liquefaction and compaction occurs every time when a next load of sand is discharged on the previous one.

For pipeline placement it's also true to say that liquefaction and compaction occurs after first sedimentation under the influence of a newly placed sand-water mixture (see 6.4, 8.4 and 9.4), but very likely, this load is much less intensive. A much stronger compaction occurs anyway for well permeable sand, through repeated loading, than for less permeable sand. Silty sand would then also obtain a low relative density when bottom dumped.

According to MITCHELL [34] the pipeline nozzle can be constructed in such a way, that the out flowing mixture loads the already present sand with such a shear strength that some compaction occurs.

Perhaps higher densities can also be achieved through higher effective stresses by applying underpressure beneath the settling surface. This has been proved under certain circumstances (small specific sand production rate, very steep slope) with small scale tests [35]. One can also think of a combination of underpressure and vibration compaction.

9.5.4 *Measures during the sand winning process*

The occurrence of flow slides during the sand winning process can result in unacceptable regression of the edge of the borrow area. During the suction or the cutter dredging in liquefiable sand, the third condition of the development of a flow slide is satisfied because it is always a matter of a relative fast exerted change. The development of a flow slide can only be prevented by remaining under the critical boundary of the second condition. The applied change itself is often characterized by a steep front slope. Recommendations are aimed to limit the height of this front slope (slopes are dredged in steps – the so called „boxcut" – with a cutting thickness of maximal 2 to 4 m) and a relatively flat final sand slope (on average often 1 : 6). It can be stated that experience has taught us that when these criteria are applied flow slides generally will not occur.

CHAPTER 10

COMPACTION

10.1 Introduction

The density of hydraulically placed sand under water is generally very small. In many instances it is worthwhile considering to compact the sand. With this the following is intended:

- reduction of the porosity, through which the stress and strain characteristics of the soil improve;
- the achievement of a homogeneous foundation through which the settling differences decrease.

Without pretending to be complete, a summary is presented in this chapter of the most common compaction methods with their applications. Apart from the systems described here, there are other intermediate techniques which are successfully applied. The following methods will be reviewed here:

- Vibro compaction;
- Vibro replacement;
- compaction piles;
- surface compaction with vibrator plate or vibrator roller;
- blasting;
- dynamic consolidation.

Besides these methods, for the sake of completeness, reference is made to the possibilities of influencing the density through the process of pipeline placement or bottom discharging (see Chapter 9).

The methods are based on one of the following principles or a combination thereof.

Generation of vibrations in the soil

With this the critical value of the ratio shear-stress/effective-stress is temporarily and locally exceeded and hence a rearrangement of the sand grains in the grain skeleton occurs. Because of surface loading, this rearrangement implies a greater density.

It almost always concerns saturated sand. Densification can than only take place when the water has the opportunity to flow away. With that, excess hydrostatic pressure always develops, more so when the soil is less permeable and the seepage distance is longer. If the sand is liquefied (effective stresses no longer exist), the grain skeleton cannot transfer vibrations anymore. That's why for this method of compaction the soil must be sufficiently permeable.

Displacement of the soil present
The soil is mechanically displaced during which excess hydrostatic pressure and flowing out through the pores, takes place again resulting in a smaller pore volume and a greater density. During these applications the soil is pushed side ways; the spaces developed herewith will be backfilled with sand, gravel or broken stone.

10.2 Short description of methods and applicability

In Table 2 and Table 3, which are copied from JAWORSKI [38] in a slightly changed form, a summary can be found of the in 10.1 mentioned techniques with corresponding characteristics. These Tables can be clarified as follows:

Vibro compaction
With this method the soil is compacted by horizontal and/or vertical vibrations which are transferred into the surrounding soil by a vibrating tube. With Vibro compaction about one metre of the upper layer will not be compacted. The following systems can be distinguished:

a. System with horizontal vibrations (Vibro flotation, Rütteldruckverfahren)
   This system [39] is based on the transfer into the ground of horizontal vibrations. A tube containing a motor with an eccentric, is jetted into the soil and subsequently pulled out while vibrating, during which at intervals of 1 m or less, the vibration takes place until the soil is supposed to be sufficiently compacted. At each interval often clean sand is supplied from the surface in the created space around the tube. The fines content ($D < 60$ μm) in the soil to be compacted, must be limited to 15 %. For higher fines contents, compaction is hardly obtained. Often a spacing of 2.5 m between the compaction points is chosen for 25 kW and 3.5 m for 75 to 150 kW motors. The spacing to be maintained depends also on the grain size distribution of the soil.

b. System with vertical vibrations
   Examples: The Terra-Probe system developed in the United States and the system developed by Van Hattum and Blankevoort in the Netherlands which, among others, has been used by DOSBOUW for the storm surge barrier in the Eastern Scheldt.
   The vibration element consists of a tube with a vibrator on the top side. The lower part of the tube, the resonator, is often fitted out with vertical fins to increase the transfer of vibrations to the sand. The tube is lowered by jetting, by vibrating or a combination of both. Otherwise the mode of operation is identical to Vibro flotation.
   The Terra-Probe technique appeared often most effective in well graded sand and gravel with less than 10 % fines. Cases are known that the method was not effective in saturated sand with a $D_{50} = 150$ μm and 5 % fines. However also cases are known for

96

Table 2. Summary of compaction methods.

method	trade name or company	description	compaction point spacing (m)	compaction depth (m)	soil types			
					gravel	sand	silt	clay
Vibro compaction	Vibro flotation	horizontally vibrating tube (Vibro flot)	1.0–3.5	24	x	x	x	
	Terra-Probe	vertically vibrating tube	1.0–2.5	24	x	x		
	Van Hattum and Blankevoort	vertically vibrating tube with steel fins at the base	2.0–7.0	20	x	x	x	
	Toyemenka	combined vertically and horizontally vibrating tube	3.0–5.0	24	x	x		
Vibro replacement	Vibro flotation	displacement of backfill by inserting horizontally vibrating tube	1.0–3.0	15		x	x	x
	Compozer	displacement of backfill by inserting vertically vibrating tube	1.0–3.5	–		x	x	x
	Van Hattum and Blankevoort	displacement of backfill by inserting vertically vibrating finned tube	2.0–6.0	20		x	x	x
vibrator plate			1 × plate diameter	0.5 to 1.0 × plate diameter	x	x		
compaction piles	Franki	displacement of sand and gravel plug with drop weight	1.0–2.5	25	x	x	x	x
blasting		delayed underground blasts from light charges	3.0–9.0	–	x	x	x	
dynamic consolidation	Ménard	free fall of heavy weights	>appr. 14	appr. 18	x	x	x	x

Table 3. Compaction mechanisms.

method	induced stress	simplified loading diagram
Vibro compaction a. vertically vibrating tube	shear stress $\Delta\tau$ at tube wall	
b. horizontally vibrating tube	shear stress $\Delta\tau$ at tube wall plus deviator stress $(\Delta\sigma_H - \Delta\sigma_R)$ from impact of tube	
Vibro replacement	shear stress $\Delta\tau$ from vertically or horizontally tube variation plus deviator stress $(\Delta\sigma_H - \Delta\sigma_R)$ from soil displacement	
vibrator plate	alternating deviator stress $(\Delta\sigma_1 - \Delta\sigma_3)$ from vibrations	
compaction piles	shear stress $\Delta\tau$ from impact plus deviator stress $(\Delta\sigma_H - \Delta\sigma_R)$ from soil displacement	
blasting	shear stress $\Delta\tau$ and deviator stress $(\Delta\sigma_H - \Delta\sigma_R)$ from shear and compression waves	
dynamic consolidation	deviator stress $(\Delta\sigma_1 - \Delta\sigma_3)$ from weight impact	

which relative densities of more than 80 % were achieved in uniform sand with a D_{50} of 170 to 700 μm. The distance between the compaction points is often chosen between 1 and 2.5 m.

With the Van Hattum and Blankevoort system often satisfactory results (relative density approx. 65 %) are achieved in fine uniform sand (D_{50} = 150 to 200 μm). The fines content in the Eastern Scheldt sand amounts to approximately 4 %. From tests it appeared that satisfactory results could also be achieved with a fines content of up to 10 % [40, 41].

Extra heavy vibration elements were used here: the diameter of the resonator (including fins) was 2.1 m and the power of the vibrator 100 kW. For that reason the distance between the compaction points could be taken rather large: in that way for instance a triangular pattern was applied with sides of 4.5 m and sometimes even up to 7 m (one compaction point per 25 to 30 m²). The water depth was a maximum of 30 m and the compaction depth 6 to 10 m. The production per vibration element was approximately 100 m³/hour. The compaction vessel Mytilus, was equipped with four of these elements. With that, 3.7 million m³ of sand was compacted in 3.3 years.

Vibro replacement

Examples: Vibro flotation techniques, Van Hattum and Blankevoort and compozer piles.

The system is more or less based on the same equipment as used for Vibro compaction (both vertical and horizontal vibration is in use). After the vibration tube has been lowered to the required depth, stone, gravel or coarse sand will be supplied via the tube or along the tube as a backfill, across a height of approximately one metre. After that the fill is driven into the surrounding soil by means of this vibration tube. This process is repeated every metre up to ground level. The spacing between the compaction points is often chosen between 1 m and 3.5 m. Vibro replacement has been successfully applied in soil varying from fine sand with 10 % fines up to clayey sand with 50 % fines. With this technique stone columns are applied in cohesive soil with bearing capacities of 150 kN in soft soil and bearing capacities up to 600 kN in stiff clay and dense silt. Extensive tests were made in the Eastern Scheldt [42, 43].

Compaction piles

Example: system Franki.

This system is based on a combination of horizontal displacement and vibration caused by driving. The procedure is very much like Vibro replacement. A tube provided with a valve at the lower end or with a gravel plug in the tube at the lower end, is driven to the required depth. Then the tube is gradually filled with sand or gravel or broken stone. After filling (1 to 2 m), the tube is pulled up and the material is pushed into the ground through driving or by means of compressed air. Compaction piles are successfully applied in granular sediments, varying from silty sand to gravelly sand with 25 % fines. Distances between compaction points amount to 1 to 2.5 m with tube diameters of respectively 0.3 to 0.5 m.

In order to increase the relative density of sand in this way, for example from 30 % to 65 %, approximately 5 % of the volume of the existing soil must be displaced and filled up.

Surface compaction with vibrator plate or vibrator roller

Compaction of a sand fill, of which the surface is situated above water level, can well be carried out from that surface with a vibrator plate or a vibrator roller. However the influence of that compaction does not reach very deep: 0.5 to 1.0 times the width of the plate, up to maximal 1 to 2 m below ground level. For that reason, it can be questioned as to what extent it is possible to compact sand under water layer by layer.

Plate compaction under water has been successfully applied on gravel and dumped stone at the Eastern Scheldt storm surge barrier [40]. It appeared necessary to provide the plate with holes. Application on sand is also possible [44], but only applicable in a practical way when the sand is covered with a reasonably thick filter layer of gravel and sand, because otherwise the plate sinks in the ground as a result of the generated excess hydrostatic pressures (liquefaction).

Blasting

Soil compaction through blasting is not applied very often. Experience with this technique indicates that it can be applied in saturated sand and gravel with less than 20 % fines.

Dynamic consolidation (method Ménard)

This compaction technique is carried out in the dry by means of a drop weight of 20 to 160 tons. The weight is dropped on the ground from heights of 15 to 40 m. This techniques has also been successfully applied under water [45]. Here a heavier drop weight must be used because the effect of the weight is limited by the resistance experienced by the weight while falling through water. Internationally, the Ménard method is often applied.

10.3   Factors influencing the degree of compaction

The most important factors influencing the degree of compaction are:

- soil characteristics, especially the permeability; an indication for the permeability is the percentage of fines;
- distance between the compaction points;
- amplitude and frequency of the vibrations;
- diameter of the compaction equipment;
- compaction time: sufficient time must be taken for compaction.

10.4   Selection of the method and control of the results

When selecting the method, first the permeability and the sand density must be estimated directly after filling by means of for example a grain size distribution, and the required density must be established.

After the main outlines of the compaction system are selected it is often recommended to optimize the most important characteristics (spacing, vibration time, amplitude, etc.) with prototype tests. On the test areas an extensive soil investigation must take place in order to judge the results. Furthermore it is also recommended to carry out a soil investigation in the whole area to be compacted after the sand is placed in order to adjust the above mentioned characteristics to the local sand conditions.

After that, the compaction can take place. The first stage of the density control occurs during the compaction process itself and often consists of monitoring some parameters of the compaction equipment itself such as energy consumption (for instance to be measured by means of an ampere gauge).

A second check is also required. For that, a new soil investigation must be carried out: This check consist of a comparison of the results of the measurements in the soil before and after the compaction is carried out. The next measurements often take place:

- Dutch cone tests or SPT's or pressiometer tests or a combination of these;
- electrical resistance measurements (electrical density tests) or nuclear density tests;
- measurements on consolidation as a result of the compaction.

Some important characteristics of these measurements are indicated hereafter. With + and − the score of the characteristics is assessed:

Dutch cone test:

− continuous with depth;
+ sensitivity of the results to execution is only limited;
+ known in the Netherlands, references widely available;
− measurement of resistance of a limited horizontal column.

Standard penetration test:

− discontinuous with depth;
− result sensitive to execution;
+ internationally known, extensive information on correlation widely available.

Pressiometer test:

+ provides stress and strain parameters;
− discontinuous with depth;
− result sensitive to execution.

Electrical hydrostatic pressure gauge:

+ semi-continuous with depth.

Nuclear density gauge:

+ semi-continuous with depth;
− strong distortion of the density because of the point.

Consolidation measurements:

+ general impression of the soil compaction across the whole area;
− the density distribution with depth is not known.

CHAPTER 11

MEASUREMENT OF PROCESS PARAMETERS

11.1 Sand characteristics

In order to describe adequately the type of sand to be used, the following characteristics have to be assessed:

Mineral composition
The possible silt and organic content of sand has to be assessed. If the sand is not composed of pure quartz, the chemical composition and the density of the grains ϱ_s has to be established (for instance with a pycnometer). Standard methods are available for this.

Grain size distribution
The grain size distribution and with that also the median grain diameter D_{50} can be determined by sieving or by means of optical methods (for instance a Malvern). Standard techniques are available for this.

Fall velocity distribution
If necessary, the fall velocity can be calculated with empirical formulae based on grain size, density and if required, temperature and water viscosity (see for example [12, 15]). It is better to test a sample, preferably on the spot or in the water of origin, in a fall column for instance the visual accumulation tube (VAT) (see [12]). With this the fall velocity distribution can be measured, from which the median fall velocity W_{50} can be derived. Standard methods are developed for the use of the VAT or other types of fall columns.

11.2 Flow rate and concentration in the discharge pipe

11.2.1 *Mixture flow rate*

The mixture flow rate Q consists of a volume of water plus a volume of sand expressed in m³/s. This flow rate is often measured on board of the suction dredger which takes care of the delivery of the sand-water mixture through the pipeline.
Flow meters however can be placed at other locations in the pipeline as well. If the total production is distributed across different pipelines, then the flow per pipe must be measured or simply be determined by dividing the flow between the different pipes proportional to the cross section.
The most common type of flow meter is the electro magnetic flow meter (EMF) (see for example [16, 18]). An EMF can be mounted in the pipeline with the same internal

diameter as the delivery pipe and thus does not influence the flow. This is an advantage with respect to the conventional flow meters such as the measuring flange. The principle of measurement is based on the influencing of a magnetic field by a flowing medium. In fact the flow velocity is measured; the meter is calibrated for the corresponding flow rate. The principle of measuring is not very sensitive for variations in density.

For the use of the EMF standard methods are available.

11.2.2 *Concentration or mixture density*

The measurement of the sand concentration or density of the sand-water mixture in the delivery pipe, is still not standard practice everywhere. Most of the time the concentration or mixture density is measured on board of the dredger, together with the mixture flow rate. For this, often nuclear methods are used, based on radiation absorption. This type of gauge must first be well calibrated (see also [46]).

A simple method to determine the concentration on the fill area at the end of the delivery pipe is by taking samples (see for this 11.5).

In addition to the methods mentioned here, the mixture density can be determined by measuring the pressure head across a vertical section of the pipe, but this is not an accurate method.

11.2.3 *Sand production*

The sand production per unit of time is expressed in cubic metres of in situ placed sand or in kilograms of the solids. For a description of these definitions, reference is made to appendix A. If the mixture flow rate and the concentration are measured, the sand production is also known.

11.3 Geometry of the fill area above water level

The geometry of the fill area above water level is determined by, among others, the minimum dimensions of the sand body to be constructed and the required working space on the crest. In the event of surveying the abovewater fill area it generally applies, that this has to be carried out in a simple and quick way because of the construction activities and the danger for quicksand. When measurements really have to be done on the abovewater fill area, it is advised that at any event two persons perform (if necessary using a rope) these measurements because of safety considerations (the likely chance of sinking in quick sand in the fill area). Measuring experiences during the hydraulic placement of the Oesterdam and Philipsdam (Delta works) have, among others, led to these observations (see Fig. 62).

For a sand body the dimensions of the abovewater fill area can be established in a reasonably safe manner by determining the length along the guide bunds (if present),

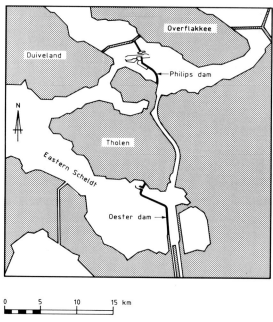

Fig. 62.   Situation of Oesterdam and Philipsdam.

and the width across the already constructed sand body behind the discharge nozzles. The measurements can be performed with a simple measuring tape of sufficient length or quicker, if necessary, with optical distance measuring equipment.

For the measuring of horizontal distances of (time dependant) phenomena on the abovewater fill area, areal photographs are a possibility. These areal photographs can be made by means of a photo camera attached to a balloon or kite. For this purpose the use of helicopters is even more accurate. The advantage of areal photographs is that they provide a total view.

Furthermore no restrictions are experienced from construction activities and no danger exists for sinking away during surveying.

The slope of the abovewater fill area can be determined using a levelling instrument (to be erected on the guide bund). Vibrations of the abovewater fill area caused by the bulldozers present, can somewhat hamper these measurements.

The two dominant forms of transport on the abovewater fill area – channels and cascades – are more difficult to measure. For this, the fill area itself has to be entered. The slope of the channels can again be determined with a levelling instrument. The beacon required for this – furnished with a bottom plate to prevent sinking away in the fill area – can be used to measure the width and the depth of the channel. The same instruments can be used for measuring the cascades. For the determination of the slope and the height of the cascade a hydrostatic level (see Fig. 63) is a quicker and more

104

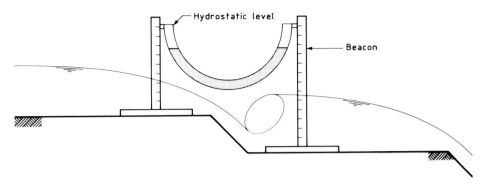

Fig. 63.   Hydrostatic level.

simple, though less accurate, alternative. The shift velocity of the cascades can be established by marking one of the steps, following this during a fixed period and subsequently measuring the displacement. For this, aerial photographs are again a good but more expensive alternative.

11.4   Mixture velocity on the abovewater fill area

The mixture velocity on the abovewater fill area can be determined by hand by measuring the velocity on the surface between two ranges by means of a stopwatch. In order to make the extent of the surface of the mixture more recognizable, floats such as ping pong balls half filled with water can be thrown in the mixture flow. This method was applied during the sand closures in the Eastern Scheldt [47].

Another method is the measuring of the height of rise by means of a robust pitot's tube, from which the velocity can be inferred as follows:

$$U = \sqrt{2gh} \qquad\qquad (22)$$

in which:

U = the mixture velocity
h = the height of rise in the tube
g = the acceleration of gravity

Due to blockage and sagging this method is not so accurate.

On the underwater fill area as well as on the abovewater fill area, the mixture velocity can be measured with an immersible electro magnetic velocity meter (EMV). The principle of measurement of the EMV is identical to that of the EMF (see 11.2.1). Because the instrument is rather robust and the measurement technique is not based on acoustic or optical principles (such as the accurate Laser-Doppler velocity meter), it is also suitable for hyperconcentrated slurry flows. In the laboratory this instrument is used for measuring sand-water mixture flows both above and under water [16, 18]. In the tilting flume, in which a uniform 0.10 to 0.15 m thick flow had established, rather accurate velocity profiles could be determined with the EMV [16, 18]. Application of this instrument on a fill area is in principle possible, but no experience has been gathered yet.

The mixture velocity of a (dilute) suspension flow under water can be measured with a small Ott type velocity meter, a method also used during the sand closures in the Eastern Scheldt [46, 47].

The mixture velocity during flow slides and the like on subaqueous water fill areas can be determined with „underwater floats" (see Chapter 12).

An instrument with which both the velocity and the concentration of the solids in a flow can be measured, is the acoustic sand transport meter (ASTM). With this instrument the sand transport rate can be measured in a rather direct manner. Application is restricted to low concentration (few percent).

If the particles are not completely in suspension, particle and flow velocity can differ. This is the case for bottom transport, during which the sand particles roll and jump across the bottom. The migration velocity of the sand particles is much smaller than the current velocity of the surrounding water. The transport rate cannot be calculated from the separate measurements of the particle concentration and the flow velocity (such as in 11.2.3). For slurry flows however, it generally applies that sand and water are completely mixed and that the sand particles are completely in suspension, because of the high degree of turbulence and the small effective fall velocity (see Chapter 6). The difference between mixture and flow velocity is then negligibly small.

For very high concentrations, such as on the fill area, the acoustic method is not useful any more. The ASTM is especially useful for accurate measurement of the sand transport rate and the sand transport distribution in rivers and channels, also for relative high concentrations (c up to approx. 5 % maximum) and relative high flow velocities. These concentrations however, can still be regarded as small compared to those in slurry flows. During the sand closures in the Eastern Scheldt the sand transport rate (suspension transport) in the closures was measured with an ASTM [24, 46].

11.5 Sand concentrations on the fill area

On a fill area above water, the sand concentration in the mixture flow (for definitions see appendix A) can simply be determined by taking samples. For this, transparent bottles can be used which are filled in the mixture flow. After the sand, and possibly the silt, has settled, the volume portion of the different components can be read from a scale on the glass. Because the deposition time of the sand is much shorter than that of the possible silt or organic fractions, a distinct stratification develops: a layer of sand, then a layer of silt and finally a layer of more or less clear water. The proportion of the different layer thicknesses, in relation to the total height from the bottom to the water surface in the bottle is an indication of the volume concentrations.

In the sand and silt layer also water is found. The water content of the sand layer in the bottle is determined by the porosity. The porosity does not vary strongly, for dense sand this is usually about 40 %, for recent deposited sand the porosity is often higher, approximately 45 to 49 %.

If the sand in the bottle is compacted by tapping lightly against the rim, it can be assumed that the water content is approximately 40 % and hence the sand content 60 %. Therefore the sand concentration amounts to:

$$c = \frac{\text{measured sand layer thickness}}{\text{total mixture height}} \, 0.6 \tag{23}$$

This method was applied on the fill area during the sand closures.

In order to prevent the local concentration distribution from being disturbed during the filling of the bottle, the „ballo"meter [47] is developed at the time of the sand closures. This instrument consists of a pipe, through which the mixture flow can pass undisturbed. By means of balls attached with an elastic band on both sides, the pipe could be closed very quickly so that an undisturbed sample could be taken. It appeared that on the fill area above water no major differences with respect to the „milk bottle" method could be observed. However, by means of the ballometer, samples can possibly also be taken in a suspension flow on the fill area under water.

A slightly more refined method to take samples is the usage of a hose pump and suction tube. By keeping the mouth piece of the suction tube at different heights or by applying simultaneously several mouth pieces each having it's own hose pump, a concentration across the vertical can be measured. A practical disadvantage of this method is, that for high concentrations and/or small hose diameters the delivery can be blocked quickly. In practice, this method is for example applied for the measuring of sand transport rates in rivers for less high concentrations. This method is also applied during the sand closure of the Krammer [24].

Especially for the benefit of laboratory research on sand-water mixture flows the conductive concentration meter (CCM) is developed by the Delft Hydraulics [16, 18]. The principle of measuring is based on the reduction of the conductivity of the water between 2 or 4 sensors caused by the presence of sand. Hence the sand-volume concentration can be derived directly from the registered signal. This principle of measuring is also applied by the electrical density measurements of the soil (see 11.6.2), where the sensors are mounted in the sounding cone.

The CCM must first be calibrated for a certain type of sand (grainsize). This can be done with a mixing tank and a suction device.

The CCM is especially suitable for high concentrations, such as encountered in slurry flows; say from 5 to 50 %. Within this range existing acoustic and/or optical concentration meters do not perform. For very low concentrations ($c < 5$ %) however, the accuracy is limited. In solid sand ($c > 50$ %, hence for soil mechanical measurements) a somewhat different calibration applies. The CCM has a small measuring volume so that rather detailed measurements are possible. This instrument, like the EMV (see 11.4), is applied in the laboratory for measurements in sand-water mixture flows above as well as beneath water level [12, 16]. Reasonably accurate concentration profiles could be determined with the CCM in the tilting flume, in which a uniform 0.10 to 0.15 m thick mixture flow was established [16]. Hence in combination with the EMV, also the sand transport rate (or sand flux-) distribution could be determined.

11.6   Measuring of sand fill densities

11.6.1   *General*

The in situ density or the porosity of a sand fill can be measured in different ways:

– electric;
– nuclear;
– with frozen samples.

The in situ density or the porosity is often only of interest if these can be compared with densities characteristic for that sand. These can be determined by taking samples and to test these in a laboratory. The following characteristics are relevant:

– minimum and maximum density;
– critical density.

Finally some conclusions can also be drawn on the density and related mechanical characteristics of the sand fill without all these measurements by means of:

– Dutch cone penetration tests.

In the following some observations will be made on each of the in situ measuring techniques and laboratory tests.

11.6.2   *Electric density measurement*

By means of electrodes mounted in a sounding probe, the electric resistance of the soil can be measured during a Dutch cone penetration test with the CCM (see 11.5). Then, with a second sounding probe, ground water samples can be sucked up at different depths in order to establish for these the electric resistance as well. Through comparison of the two resistances the porosity can be determined for the soil concerned. For that, a calibration line is required, which is not the same for each type of sand. Therefore it is recommended to collect some sand samples and to determine for these the calibration line in a laboratory.

For the distance between the two outer electrodes often one metre is chosen. In that way the porosity is measured for a volume of sand of approximately one cubic meter as an average. The disturbance introduced by the sounding probe is then relatively small. If a smaller electrode spacing is chosen, this influence is then relatively stronger. For very loose sand though, the disturbance can still be strong because the bringing in of the sounding probe can induce liquefaction in a rather large area, through which at the time of the measurement the sand will be somewhat compacted. If an electrical density meter is placed before the sand is placed (see Chapter 12), this problem will not occur. More information about the electric density measurement can among others be found in [48].

11.6.3 *Nuclear density measurement*

With probing equipment a measuring tube can be pressed in the soil in which a device is lowered for nuclear density measurements. The device consists of a gamma-radiator, a neutron-radiator and a crystal. With the crystal and the necessary electronics the intensity of radiation, diffused and weakened by the soil, is measured. The intensity of the gamma-radiation is a measure for the in situ density of the soil; the intensity of the neutron-radiation is a measure for the volume of porewater. The latter is especially of interest for saturated soil.

With this equipment the average density is measured of a volume of approximately one cubic decimetre, at least for the usual radiation intensities. The influence of the disturbance caused by the insertion of the probe is thus considerable. Very loose packing conditions are therefore seldom observed. Higher radiation intensities entail safety problems. More information can be found in [49].

11.6.4 *Density of frozen samples*

At the moment a method is developed for which the undisturbed sand soil is frozen, after which samples are obtained through core drilling. The samples are not defrosted until placed in the laboratory test apparatus by which a number of important original characteristics, among others the density, are preserved. Therefore it must be possible to determine the in situ density rather accurately. In first instance the development is aimed at the soil close beneath the surface.

11.6.5 *Minimum and maximum density*

The in situ density can be compared with the minimum and maximum density. That gives the so called relative density.

The minimum density is the smallest density which can be achieved by a grain structure in dry conditions. The porosity n is then maximum: n_{max}. When the porosity is equal to that, then the relative density is zero percent.

The maximum density is the greatest density which a grain structure can achieve in dry conditions. Then the porosity is minimum: n_{min}. When the porosity is equal to that, then the relative density is one hundred percent.

Situations are known for which the in situ density is even smaller than the minimum density as determined in the laboratory. Most of the time though, the value varies between these two limits. Two definitions are in use for the relative density D_r: one is based on the porosity n, the other is based on the void ratio e.

For the relative density based on the porosity n the following applies:

$$D_{rn} = \frac{n_{max} - n}{n_{min} - n} \qquad (24)$$

And for the density based on the void ratio $e = n/(1 - n)$:

$$D_{re} = \frac{e_{max} - e}{e_{min} - e} \qquad (25)$$

For the same situation D_r provides slightly higher values for the last definition than for the first definition. The maximum difference between both definitions is approximately 7 %.

The relative density is often used as an indication of, for example the sensitivity to liquefaction, the strength of the sand during sudden deformations, the dredgeability of the sand etc. Because the value can rarely be determined accurately, the relative density appears a rather vague parameter.

The minimum and maximum density can be determined in a laboratory in a simple manner. Unfortunately there are no internationally accepted uniform procedures. Also because of this reason it is not possible to draw definite conclusions from available values of the relative density.

11.6.6   *Critical density*

Dry critical density
A more distinct measure than the relative density is the „dry" critical density for the in 11.6.5 mentioned sand characteristics. That is the relative density for which dry or drained sand does not exactly show a volume increase (dilatation, see also 11.2) nor exactly a volume decrease (contraction) under the influence of shear deformation. The determination of these requires the performance of a few triaxial tests or plain shear tests according to certain procedures. The critical density is a function of the isotropic tension. If the critical density of the sand at different depths has to be known, then a few test series must be performed.

Wet critical density
For the sensitivity to liquefaction under static loading and thus for the sensitivity to flow slides, an even more distinct measure exists; namely the „wet" critical density. This requires triaxial tests according to a slightly different special procedure.
Both procedures are among others described in [50]. The concept of „critical density" is an international accepted concept. With that the dry density is meant. The addition „dry" is introduced by LINDENBERG and KONING [50] together with the concept of „wet critical density".

Critical state density and steady state density

Also widely used are „critical state density" and „steady state density" with nearly the same definition. This density (these densities) lays (lay) in between the dry critical density and the wet critical density.

11.6.7. *Dutch cone penetration tests*

The results of Dutch cone penetration tests are also strongly influenced by the (relative) density of the sand. Because of that, it is possible to use these tests for extrapolating densities encountered in a certain location to other locations. The reliability of the extrapolation depends strongly on the homogeneity of the soil structure. With sufficient (geologically founded) knowledge and experience conclusions are also possible on the basis of Dutch cone penetration tests only.

CHAPTER 12

SPECIAL MEASUREMENTS

12.1 Measurements of flow slides

12.1.1 *General*

If, during the construction of a sand body under water, the possible occurrence of a flow slide ought to be measured and the size of this ought to be determined, various special measuring techniques can be applied. Such special measuring techniques were applied during the placing of the sand dam for the closure of the Slaak as part of the Delta works (the Netherlands) in August and September 1986. These measurements are extensively reported in [51]. Some results are described in [52]. A short description of the measurements carried out at that time is given here, followed by a discussion of the usefulness of the most relevant measuring techniques.

For the measurements in the Slaak a small platform was installed, consisting of three driven piles connected by two catwalk bridges. The platform stood in the centre line of the future dam, close to the middle of the channel where the water was originally 7 m deep. Measurements took place during about four periods of twenty four hours. At the end of that period, the sand surface had practically reached the water surface. Figure 64

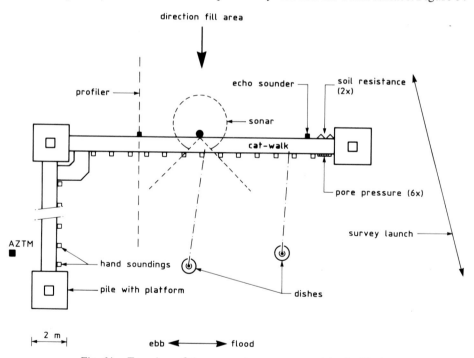

Fig. 64. Top view of the measuring arrangement in the Slaak.

provides a top view of the platform with an indication of the location of all measuring instruments.

The most relevant methods for measuring are echo-sounding surveys (inclusive the profiler), the side-scan sonar, the pore pressure meters, the density meters and the „dishes".

12.1.2 Echo-sounding surveys

Frequent echo-soundings combined with an accurate positioning system provide important information on the changes of a slope. From the changes of the slope surface between two subsequent surveys, one can often conclude whether a flow slide has taken place and what was the extent in a certain direction. It will be necessary that the survey is done as much as possible in the same range. Otherwise differences in profile as a result of a flow slide can often not be distinguished from differences in profile as a result of location differences. Furthermore the frequency of the measurements must not be much less than once in half an hour, in order to distinguish the sudden change as a result of a flow slide from the gradual change as a result of „normal" sedimentation. These problems can be overcome by using a profiler; this is a rotating echo-sounder which measures in a vertical plane and registers a profile every few minutes from a fixed point. Needless to say that from the numerous surveyed profiles eventually only a few are required.

An outstanding feature of a flow slide however is the speed with which this takes place: within one minute the bottom can rise or sink one meter or more. The best way to observe this is with an echo-sounder which registers from a fixed point. A profiler provides much more information but also it's reach is limited: one section with a length of not more than some dozens of metres.

12.1.3 Side scan sonar

A side scan sonar is able to survey within 20 seconds a considerable area with a diameter of some dozens of metres. The type of information however, differs from that of a common echo-sounder or from that of the profiler: the local depth is not indicated but the degree of reflection of the sound waves. The surface of the bottom is made „visible". One can see whether locally the slope is much steeper than elsewhere. When the same surface is surveyed every twenty seconds from the same spot, then the possible „walking" of a steep slope can be observed (see Fig. 65).

12.1.4 Pore pressure meters

Echo-sounding nor side scan sonar surveys provide information on the thickness of the sand volume which takes part in the flow slide. Pore pressure meters, inserted at different depths (for example every metre) do provide this information. These gauges can for example be mounted on sounding probes. From the plot in time of the hydrostatic pressure one may furthermore conclude whether it is a matter of a flow slide. This

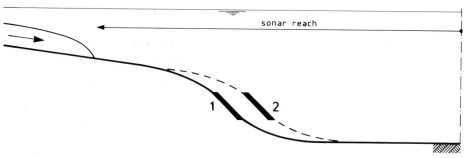

a. sand-water mixture flows from slope : slope section moves forward

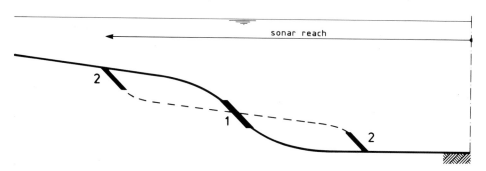

b. flow slide : two sections of steep slope move in opposite directions

Fig. 65.   Visible sections of a steep slope on side scan sonar at point of time 1 and at point of time
2 (20 s after point of time 1).

namely is characterized by a rise of the hydrostatic pressure in less than one second and
a gradual fall during (order of magnitude) one minute. Depending on the depth of the
gauge and the thickness of the flow the hydrostatic pressure may remain constant after
the sudden increase. One can derive the thickness of the flow from the duration period
and the height of the excess pore pressure.

12.1.5   *Density meters*

After initial deposition of sand from a sand-water mixture, the density of the sand
volume is usually very small. During a flow slide the density normally increases
noticeable. This process may be observed by setting up electric or nuclear density-
meters at the location of the future sand body (see Chapter 11).

12.1.6   *„Dishes" or „underwater floats"*

In order to register whether and how the surface of the underwater bottom moved
during a flow slide, a special instrument has been developed for the measurements in

the Slaak (see Fig. 66). This instrument consists of a dish made of wood or steel having such density that the dish sinks in water but floats in a sand-water mixture. The dish is connected to the platform by a fishing line provided with a measuring system. A great part of the line is coiled on a fishing reel which is attached to the platform. Approximately 10 dish movements were observed in the Slaak over a distance of 1 to 12 m with velocities up to 0.5 m/s. Two examples are drawn in Figure 67. These dish movements took place at the same moment as the registering of the sand-water mixture with the side scan sonar or other method of observation.

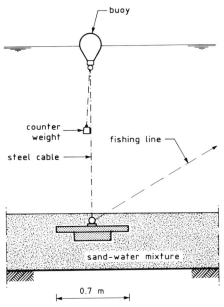

Fig. 66.   Dish for registering the movement of ground surface.

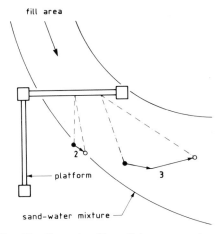

Fig. 67.   Example of two dish movements.

115

CHAPTER 13

SUMMARY OF THE THEORY FOR PRACTICAL USE

13.1 General

The last few years considerable knowledge has been acquired on the process phenomena of building with sand in water through model research, theory development and measurements in the field. This knowledge is qualitative and quantitative, but not sufficiently developed to carry out detailed calculations. The accuracy with which many of the parameters can be calculated is not more than a factor 1.5 or 2. In some cases only the order of magnitude can be determined.

13.2 Slope formation in the borrow pit

In many cases the slope developed in the borrow pit, will not be much flatter than the natural slope, for example a slope of 1 : 2 to 1 : 3 (see 3.1). In two situations a much flatter slope can develop:

– when layers of soft clay or peat occur within or below the borrow area;
– when the sand is very loose.

In the first case the slopes can be predicted with stability calculations, provided a reasonable assessment of the friction characteristics of these soft layers can be made. The soil investigation must be geared to that (see 3.2 and 3.3).
In the second case flow slides can develop. In 9.3 it is explained how the risk of flow slide occurrence can be determined. For sands in the Netherlands the average slope gradient developed after a flow slide, is often 1 : 5 to 1 : 15. During dredging, most of the time, flow slides can be prevented by cutting flat slopes in a careful manner (in many situations in the Netherlands for example a slope of 1 : 6) in rather thin layers (2 to 4 m).

13.3 Falling of a sand-water mixture through water

Three methods of discharging can be distinguished during which the mixture is falling free through the water column (see 2.1, 4.1 and Fig. 32). During pipeline placement (single point discharging) in an underwater fill area the sand-water mixture falls through water like a conical shaped jet or plume; during curtain-like discharging like a vertical plane shaped jet or plume (curtain); during lump-like discharging like a more or less cylindrical shaped lump. For lump-like discharging no calculation method is available. However it is likely that this rarely occurs, because, when using a split dumping barge the lump has most probably the shape of a long thin wisp or curtain, so that also in that case it is a matter of curtain-like discharging (see 4.1.1).

116

With a jet the mixture movement is especially determined by the initial velocity; with a plume the movement is especially determined by gravitation (the mixture is heavier than the surrounding water). Possible cross current plays a role in both cases.

Computer programs exist with which the following quantities can be calculated as a function of the depth under the discharge nozzle, or under the valves, or whereever the mixture may be discharged from:

- the trajectory of the jet (the horizontal distance to the point of discharge in connection with the possible horizontal initial velocity and/or the cross current);
- concentration c in the centre of the jet (c always decreases due to entrainment of water into the jet);
- diameter d or width b of the jet (this increases most of the time through entrainment through decreasing velocity).

The Figures 3, 5, 6 and 34 give an impression of these parameters for a usual case. The concentration c remains constant up to a distance of $8d_0$, respectively $8b_0$, in which d_0 and b_0 are the values at the point of discharge.

13.4   Crater formation

For most cases with more or less vertical falling stationary jets, the crater depth y_s developing as a result of a clear water jet, can be determined with formula (4) of BREUSERS or formula (5) of RAJARATNAM (see also Figures 37 and 40). The crater depth y_s has to be reduced to y using Figure 41 in order to find the crater depth developed as a result of a sand-water mixture jet. For round (conical) shaped jets formula (7) of HEEZEN and VAN DER STAP can also be used (see Fig. 42). This fomula gives y directly; this value does not have to be reduced.

As input quantities for the above mentioned formulae the jet characteristics near the pipe nozzle or the hopper has to be used. However, the formulae are not valid if the falling height is too large or too small (see 5.1.1 and 5.1.3).

For larger falling heights, and also when the jet describes a strongly curved trajectory, the formula of HEEZEN and VAN DER STAP may be used, provided that, the jet characteristics just above the bottom are chosen as input quantities.

In all cases the width or diameter of the crater can be inferred from the depth by means of formula (6).

For non-stationary situations, such as hopper- or dump-barge placement and such as single point discharging while shifting the nozzle in a quick way, the crater width or crater diameter is presumably more or less equal to that developed during stationary situations. The crater depth will then be smaller. How much smaller cannot be predicted without model tests, but an upper limit can be estimated according to the method indicated at the end of 5.2.

In general one can say that the crater dimensions are especially influenced by the pipeline diameter, the split width, the valve dimensions and the initial velocity of the mixture. The larger these are and hence the greater the mixture flow, the larger the crater dimensions. Besides that, also the grain size and the falling height plays a role: finer sand and greater falling height lead to greater dimensions. See also the calculations for the analyses of a practical case (21.3.2, table 15) and those on the case histories (especially appendix B, tables B2, B4 and B6). A spray head which takes care of a strong reduction of the discharge velocity can also take care of a reduction of the size of the crater, as follows from the calculations of the case history „trench filling Beerkanaal" (zie 20.7.6).

For the overflow of the mixture from the crater, the mixture flow rate Q, the concentration c and the width B at which the mixture spreads is of importance. The first two quantities are presumably equal to those of the radius at bottom level. Little is known about the spreading width B. As a maximum the circumference of the crater can be taken. However, most likely, the mixture concentrates in a channel. The width can be estimated with the formulae (11) or (12) (see the Figures 44 and 45). That width can probably be regarded as a lower minimum. It may be assumed that the width remains more or less constant between crater edge and point of deposition.

With the formulae (13) and (14) the specific flow rate q and the specific sand production rate s can subsequently be calculated.

13.5   Sedimentation on a fill area above water

The length along which deposition from the mixture occurs on a horizontal bottom is approximately equal to the sedimentation length L. This follows from Figure 48. If L is longer than the fill area above water, than L defines the average slope. However, most of the time, L is much shorter. Then sand settles at an average slope being approximately equal to the equilibrium slope. This can be determined with formula (15) (see Fig. 53). Generally speaking coarse sand results in steeper slopes than fine sand. Another important parameter in this formula is the specific flow rate q and thus the width according to which mixture is distributed. The greater the distribution, the shorter the sedimentation length and the steeper the slope.

13.6   Sedimentation on a fill area under water

Depending on the question whether flow slides occur directly after the sand has settled on the slope of the fill area under water for the first time, two or three types of sedimentation can be distinguished (see Fig. 54). In order to predict which type can be expected, the sedimentation length has to be estimated (see Fig. 48), as well as the quantity h_{cr} (see formula (17) and Fig. 55). With that the dimensionless parameters L^* and H^* can be determined with the formulae (18) and (20). The shape of the slope is mainly determined by flow slides if $H^* > 1$ and $L^* < 1$. Also N^* has to be 1 (formula (19)), but that is almost always the case during the initial deposition on an underwater fill area (see 8.4). The slopes which develop can be found using formula (21) (see Fig. 56). The formula is applicable if no flow slides develop. The moment these do occur, the slopes often become flatter. That can be seen on the points in the Figures related to observations in the field. It is shown that coarse sand and low specific productions are required to achieve a steep slope.

More information on the occurrence of flow slides during and after the construction of sand bodies can be found in Chapter 9. Very important is the developed porosity, or relative density D_r. For placement under water, these are generally very small ($D_r = 0.2$ to 0.4), so that $N^* > 1$. Data exist which indicate that during bottom discharging higher densities can be achieved than during single point discharging. However, whether this is generally applicable remains to be seen (see 9.5.3). In Chapter 10 which possibilities are available to increase the density are explained.

practise

CHAPTER 14

EXTERNAL CONDITIONS DURING CONSTRUCTION

14.1 Subdivision into groups

Both for the design as well as for the construction of a project nature dictates a number of project conditions. In this chapter only conditions with respect to the construction will be dealt with. These can vary widely and restrict or even prohibit the application of certain work methods. Roughly these conditions can be split into three groups:

group a. those which determine the use and efficiency of equipment, briefly summarized under the concept of workability;

group b. those arising from a change in the hydraulics and morphology of the area, occurring as a result of the works; this influence can be of a temporary or permanent nature;

group c. those arising from environmental considerations; also this effect can be of a temporary or permanent nature.

Each of these groups of conditions are discussed in the following paragraphs.

14.2 Workability

The workability of floating equipment on the project location is influenced by:

– climatic factors, such as fog, frost, ice occurrence, storm, sea state and swell;
– hydraulic factors such as current velocities and variations, current gradients and water level variations;
– other factors such as reducing shipping hindrance.

Of these factors the climatic ones are the most difficult to predict. Their occurrence can only be described in a statistical way and sufficient data must be available for this. Based on the statistical processing a conclusion can be made on the probability of occurrence or exceedance of certain conditions. The deviation can be considerable, so that the conditions occurring in practise divert from the average expectation. As the execution of a project takes longer, the probability of approaching the average frequency of exceedance will increase. In practical terms this means that the loss of time on a project of a short duration, as a result of extreme conditions such as a gale, cannot be regained during the works.
For the estimate of the workability of a dredger and auxiliary equipment different situations can be distinguished:

– operational conditions: the equipment is not hindered;

- marginal conditions: the production decreases and there is a chance of damage (which in general can only be repaired when conditions improve);
- unworkable: the equipment lays idle;
- critical conditions: the equipment must survive or seek refuge; this, among others, occurs during gale or hurricane warning; seeking refuge must take place in sufficient time, hence on a forecast and therefore on occasion retreat may take place unnecessarily.

Waves

Often wave analyses are based on data which apply to a large sea area. In order to make these applicable to the work location, the following circumstances are of importance:

- protection against wave propagation from certain directions by islands, coasts, shallows and breakwaters;
- shoaling, breaking of waves, refraction and diffraction;
- seasonal influences (monsoon winds);
- daily variations (change of wind direction from shore to offshore and vice versa, especially in desert type coastal areas);
- flow of the water.

The ship movements are strongly influenced by the relation between ship length and wave length. When the wave length is much shorter than the ship length the vessel will „sense" very little. When the wave length is much larger than the ship length the vessel tends to follow the wave. Large pitching movements develop when the wave length equals 1.5 to 2 times the ship length. Therefore it is important to distinguish between sea state and swell.

sea: waves are generated by local winds (wave periods are usually less than 5 s);
swell: waves are generated at a great distance at an earlier time (wave periods exceeding 5 to 6 s).

The response of a ship to wave action depends on the dimensions of the ship, the orientation of the ship with respect to the wave direction (sometimes this cannot be chosen optimally) and the anchoring system. Ships anchored on wires (slack system) possess a lower natural frequency than ships anchored on spuds (stiff system). Depending on the peak periods of the wave spectrum of a certain wind field, the response will be larger or smaller for a certain anchoring system.

Sometimes it is not the movement of the main equipment which determines the workability, but other parts of the dredging spread. For example for a dredge spread this can be the spuds, the cutter or the floating pipeline. Mooring alongside vessels, for the loading of barges for example, is often impossible during swell.

Wind and current loads

Not only the movements, but also the forces exerted on equipment or parts thereof can

Sand discharging of hopper dredger via floating pipeline making use of a cutter suction dredger as booster station.

be the controlling factors, such as the anchorage system, wires or winches, or the presence of bad anchoring ground. These forces can be the result of a direct wind load on the abovewater part of the spread or a current load on the submerged part. In the event of floating ice pushed up as a result of wind or current action against the equipment, these forces appear to be considerably larger because of the extra resistance of the current above or below the ice surface.

The general formula for the force exerted by wind or current action against a body is as follows:

$$F = 0.5 \varrho v^2 C_s A$$

In which:

ϱ = density of the surrounding medium (air, water) (kg/m^3)
v = velocity difference between body and medium (m/s)
C_s = pressure coefficient, dependant on the shape
A = exposed surface perpendicular to current (m^2)

The influence of the pressure coefficient is very important. For its magnitude in various situations reference is made to the literature.

The hydraulic factors, being variable as well, can be generally better determined than the climatic factors. This mainly applies to river discharge characteristics and tidal influences. There where the climate has a strongly variable influence, such as bandjirs (sudden strong rain shower discharges) or water level set up (raise of water level as a result of strong wind), a reliability analysis is required to be able to assess the risk of loss of production time.

14.3 Hydraulic and morphological changes

When a sand fill is built up in an area being subject to currents, a change of the hydraulic circumstances occurs during the project execution. This has influence on the distribution of the current velocity and sometimes on the magnitude of the current. As a result, erosion of the sand fill under construction can develop and also in the wider surroundings variations may occur in the erosion and sedimentation balance. The variation itself changes during the progress of the works but obviously this is an interim phase which ultimately leads to the state of equilibrium of the new situation.

In those cases where the water depth is limited, or the total flow cross-section is restricted, the dredger with it's auxiliary equipment (floating pipeline) can be a large obstacle in the flow. Because of this, apart from current forces on the dredger, also extra erosion forces on the bottom and embankments can develop.

Changes also have to be taken into account as a result of the sand winning. With a deep borrow pit a large influence can be expected on the sedimentation pattern in the borrow area. This may only be restored when the borrow pit has mostly refilled itself again. At the same time there is the risk of an impermeable layer being penetrated by dredging which can cause disturbance of the groundwater flow. This not only is of consequence for the hydrostatic pressures in the bottom layers, but may also be undesirable because of environmental reasons such as the intrusion of salt water, pollution or stratification of the water column (see 14.4).

When dredging takes place in a small area, this will have a minor effect on the waves and the flow even for a deep borrow pit.

When making a relatively shallow and extended winning area or when dredging a shallow sand bank the flow can be influenced noticeably and also a change of the wave motion is possible.

14.4 Environmental aspects

For the protection of the community, conditions have to be defined both for the design of a project as well as for the implementation. In principle the community consists of three components: the flora, the fauna and the human, which furthermore are interrelated. Which of these factors will be affected most, depends on the nature of the area and on the purpose of the works. For example the hydraulic placement of a sand island in a natural area for the preservation of a bird sanctuary defines a different set of conditions when compared to the hydraulic placement of sand for a ship's berth in an industrial area.

The considerations can be summarized as a number of main items, namely:

- aspects which are a result of the project itself and which are of a permanent nature;
- aspects which occur only during the execution of the works;
- aspects which occur as a result of the project execution and are of a permanent nature.

Of the above aspects a change in the hydraulics can cause a permanent change in the geomorphology. The aspects related to the method of execution provide in many occasions a certain freedom of choice, often however with financial consequences. The following factors play an important role with respect to the development of a sand fill in water:

- sandwinning
  - By the removal of a relative thin surface layer over a wide area (hopper suction dredger) a great deal of sea-floor life will be destroyed. Because the natural circumstances will alter very little this community will re-colonize reasonably quickly starting at the edges.
  - During the construction of a deep borrow pit concentrated in one location, the disturbance is also concentrated but this may result in a rather high turbidity of the surroundings. In a later stage the borrow pit will then act as a silt trap for an extended period of time and as a result a different sea floor life will be established. The penetration over a large depth of the horizontal soil stratification may lead to a change in the groundwater flow.
  - The winning of sand always causes some turbidity. The extent depends on the fines content (silt) and the type of dredger. This may result in reduction of light penetration in the water, which may hinder plant growth and disturb crustatea living on the bottom (blockage of food intake- and digestion organs).
- loading of barges
  - During the loading of barges alongside a dredger in a borrow pit or during transfer of a hydraulic mixture from a pipeline into a barge fines will be washed overboard. Depending on the nature of the current and the water depth these fines can settle over a wide area and hinder plant growth on the surrounding sea-floor.
- fill area
  - During bottom discharging of sand from hoppers fairly high turbidity of the surroundings at the discharge location will be caused. In addition, this distribution of silt will deposit as a layer on the bottom and plants across a rather large area. Its extent will depend on the water depth, currents and fines content.
  - Also during under water pipeline discharging towards the bottom, spreading of fines occurs, but to a much lesser extent than during bottom discharging.

The extent and effect that the foregoing phenomena have depends also on the vulnerability of the area and the extent of pollution and contamination of the sea floor.

Finally, in some cases, there may be question of direct disturbance of the human environment. This can for example be visual hindrance or inconvenience caused by noise in both recreational and domestic areas.

If there is clearly seasonal influence (eg. breeding or growth period, holiday season etc.) the disturbance may be reduced to acceptable standards by including time restrictions in the dredging programme.

CHAPTER 15

STRUCTURE TYPES, FUNCTIONS AND REQUIREMENTS

15.1 Review of structure types

Functions
Sand which forms part of a structure and which has to be placed in or below water, can fulfill different functions, such as:

- formation of a protection or isolation layer;
- providing of ballast weight;
- providing vertical support or load distribution;
- providing of horizontal soil pressure;
- providing of drainage capacity;
- filling voids.

Naturally these functions also occur in combinations.

Specifications
Depending on the above mentioned applications a set of specifications of a diverse nature will be formulated for the sand to be placed.
Moreover will this function be of a permanent nature or does it only have to be maintained during a certain period (as a temporary structure)? This also leads to a set of specifications.
Finally, wrong or unjudicious applications would lead to undesired side effects. These could occur both during the building stage (method of construction) as well in the final product. With that, a third set of specifications can be formulated.

Structure types
Structure and/or projects types can be sub-divided as follows:

- sand backfills and covering of pipelines, whether or not in pre-dredged trenches;
- sand backfills around tunnels and large ducts which are positioned in pre-dredged trenches by immersion;
- sand backfills and embankments behind quay walls formed by *L*-shaped walls, caissons or stone dams;
- embankments or trench backfills (soil improvements) using sand for the formation of aprons, berms or plateaus as a sub-structure for dams, islands or embankments for roads and rail roads;
- building up abovewater level of sand fills constructed under water, for example as a further development of the sand placement below water level refered to above.

Each of these structure types will be considered in the following.

Reclamation in Singapore, Empire dock.

15.2 Sand fills and covering of pipelines

Pipelines for the transport of oil, gas or other matters are sometimes placed on the river or sea bottom, but mostly inside a trench in the bottom. Often these pipelines have to be protected and sometimes also be isolated by means of a sand fill. Hence this body of sand will be an embankment around the pipeline when positioned on the bottom, or a complete or partial backfill when positioned in a trench. For reasons of erosion-durability and for protection against damage from anchors, sometimes a protective layer of broken stone or coarse gravel is placed on top of the sand.

Sand functions
The primary function is to provide protection against current and wave attack, some-times also to provide thermal insulation. In the event that a layer of stone takes over the protective function, the sand serves as backfill of voids or acts as a filter between the bottom material and the stone or gravel.

Side effects

After being placed, the pipeline will only be supported locally on the bottom or in the trench. The sand backfill will not take over this support because the sand is not compacted and, for large pipe diameters, the open space is not completely filled. Hence the vertical load must be taken over by the pipe itself and from this a maximum acceptable length follows for the free span. Placing of the pipeline on supporting structures which collapse after a certain force is exceeded may be a solution for this.

Irregular building up of sand on both sides of the pipeline, especially during placement, will cause horizontal loads especially for large pipeline diameters. Therefore even distribution of fill within certain tolerances is desirable.

Especially with respect to less permeable sands (fine, silty) a probability of liquefaction exists as a result of external loads. These loads may result for example from water movement (waves, shipping), water hammers in the pipeline and earthquakes. If the liquefaction occurs over a sufficient length then the pipeline may become buoyant in this heavy fluid.

Specifications

Depending on their relevance, specifications can be necessary for:

- layer thickness, in connection with protection or insulation value and filter function;
- grain size distribution, in connection with erosion durability, isolation value and risk of liquefaction;
- tolerances, in connection with loads during and after sand placement, disturbance of the bottom position and placement of a cover layer.

15.3   Sand backfills around tunnels and large ducts submerged in pre-dredged trenches

Immersed tunnels, which can have either a rectangular cross-section or a circular cross section, are constructed by placing elements in pre-dredged trenches after which these are coupled. Such tunnels are used for road or railway, as a transit for pipelines and as cooling water intakes and outlet ducts. The differences with 15.2 are the much larger dimensions also requiring a much larger trench and backfill volume.

The elements are generally positioned on temporary supports, after which the sand backfill in the spaces under and along the sides of the tunnel takes over the support. The backfill underneath a rectangular segment, where an open space prior to filling can have a width of 25 to 50 m, a length of 100 to 150 m and a maximum height of 1 m is a very special and meticulous operation.

In order to protect the roof of the tunnel, a cover layer of sand or sand and stone is placed, which at the same time refills the dredged trench up to the original bottom profile of the river.

Function of the sand

The primary function of the sand under and beside the tunnel is to provide support. On top of the tunnel the function is to provide protection and backfill of the dredged space. It may be also noted that the sand underneath the tunnel transfers the load to the sub-soil. If the subsoil is too soft or sensitive to consolidation, then soil improvement must be applied, hence this involves the replacing of soft layers with sand.
This sand also has a support function.

Side effects

If the tunnel is constructed in a river with a large sediment transport then there will be a large risk that silt settles in the deep cuttings during the construction period. Silt deposits, especially underneath the tunnel, can have a detrimental influence on the quality of the sand foundation.
Because the sand underneath the tunnel may become denser as a result of intermittent loading, settlement must be taken into account after the tunnel element is placed on the sand bed. Consequently a considerable thickness of the sand layer may require soil improvement by compaction beforehand.
As long as the tunnel element rests on its temporary supports the allowable ballast weight is limited. Relatively small forces caused by underflowing or backfilling with sand may result in unacceptable displacements.

Specifications

With respect to construction considerations the following specifications have to be considered:

- subsoil, in connection with the foundation method and soil improvement;
- sand gradation, in connection with bearing capacity, underflow process and hydro-static pressures;
- silt enclosures, in connection with settlements and hydrostatic pressures;
- thickness and nature of sand layer underneath the tunnel, in connection with settlements;
- layer thickness and gradation of the protective layer on top of the tunnel;
- stability of the cover layer against erosion by river bottom processes.

15.4   Backfilling and embankments behind quay wall structures

Various types of structures make up the transition between land and water. In many circumstances this transition must be as steep as possible: for instance harbour quays are vertical, and embankments of built out areas have slopes of 1 : 1 to 1 : 3. The slope gradient and the type of structure are closely related to each other. In principle there are two possibilities:

- make the quay wall structure first, for example as a caisson, a quarry stone embankment or as a stack of sand sausages (see appendix C) and backfill the structure with sand;

132

– build the structure out with sand until the desired dimensions of the area are obtained, finish off the profile of the sand slope and cover the slope with a shore protection.

With the last method the structure can never be steeper than the underwater slope achieved by shaping and which can subsequently be maintained for some time. The quarry stone embankment can never be made steeper with dumping than 1 : 1.25.
An intermediate shape can be realized by making the embankment in layers. This is realized by dumping rock bunds ahead of each layer of sand (see Fig. 68). Hence the rock bunds are stacked on each other whereas each embankment is placed slightly backwards of the previous one and rests partly on the underlying sand backfill. Owing to this stacking arrangement the resulting slope gradient is slightly more gentle than the natural angle of repose of the stone material.

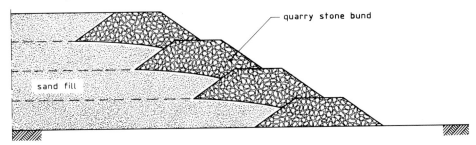

Fig. 68.   Method of building up a sand fill using rock bunds.

Functions of the sand
Depending on the structure the sand has different functions:

– behind sheetpile wall structures:
  lateral supporting force, support for backward anchorage, backfill of voids and foundation for the overlying areas;
– for caissons:
  within the caisson   – ballast;
  behind the caisson   – backfill of open space and area
                                  foundation;

– for L-shaped walls:
  on top of the bottom plate   – ballast;
  behind the bottom plate   – area foundation;
  underneath the bottom plate   – support of the sub-structure consisting for
                                  example of a sheetwall structure or slope
                                  protection;

– behind stone embankments:
  one embankment   – backfill and area foundation;
  stacking of various embankments – also support for the embankments;
– with slope protection on
  build out sand slopes: – foundation for the protection.

133

Side effects

In all these cases the sand is a surcharge both during construction and in the final works which has to be allowed for in the design. Extreme loadings may occur when (during placement or in the final stage) a sand slide or liquefaction occurs as a result of the execution or earthquakes.

Placing of sand in the frequently long narrow spaces behind quay walls results in the possibility of size sorting. As a result silt and fines will concentrate in certain areas which has a detrimental effect on the bearing capacity and enhances the risk of instability.

Water level variations of surface water as well as of the groundwater may generate excess hydrostatic pressure in the sand. Owing to this sand grains may migrate through open joints in the structure, thus affecting the function of the sand. The grain size distribution of sand, the sand density of the structure and the permeability of the sand are interrelated.

Stacked stone embankments with backfilling in layers are prone to the risk of sand migration. This migration may develop because of:

- embankment material which is too permeable;
- lenticular occurrence of sand between the embankments.

Requirements

The requirements for sand depend on the type of structure and concern:

- grading, in relation to the bearing capacity, settlements, stability and migration;
- construction method, in relation to the occurrence of silt concentrations, uneven loadings and risks of slip failures;
- density, in relation to bearing capacity, settlements and risk of liquefaction.

15.5 Underwater embankments or trench backfill using sand

Hydraulic structures are often founded on an artificially placed sand fill. Depending on the shape of the sand fill this construction can be an apron, a berm or a plateau. The sand has to support the super-structure, for example a dam, an island, a caisson structure or a (rail)road. These sub-structure type of sand fills in water, occur in various situations such as an island in open sea, a dam through an estuary or a road through a lake.

In some cases the sand fill will not be built up to the water level but finished some meters below. In other cases the under-water-part may be continued with a sand layer above water (see also 15.6). Sometimes the sand fill itself is the construction objective, such as a sand embankment for reduction of waves or an artificial bird's island.

Depending on the external boundary conditions and the nature of the construction, the side slope of the sand body has to be protected against erosion by current or wave attack. Trench backfills as soil improvement for the sub-structure of hydraulic structures also belong to this category. Such a trench backfill can be necessary in a situation where the soil has a low bearing capacity and a high groundwater level. Soil improvement will then be obtained by replacing the weak layer with sand. The sandbody forms the backfill of a pre-dredged channel and the channel slope acts as the enclosure for the sand. However, during discharging the sand-water mixture can also flow over the edge of the channel.

Functions of the sand
In general, the function of the sand is to provide vertical support. Sometimes the distribution of the surcharge is also of importance owing to the transfer of the surcharge to the sub-soil. Finally it is often a matter of backfilling of voids in order to raise the foundation of the super-structure to a higher level.

Side effects
The most important side effects for these generally rather large sand bodies, are the stability, the sensitivity to settlement and the weight of the sand body itself.
With respect to the stability, the probability of flow slide occurrences for slopes which are too steep is of importance as well as the probability of liquefaction as a result of possible cyclic loading.
The self weight plays a role when founding on soil with a low bearing capacity, also for soil improvement through trench backfilling. In the latter case the side slopes of the channel as well as the subsoil below the sand backfill may fail as a result of surcharging with sand. Monitoring of the hydrostatic pressures in the soil and construction in phases by building up in layers with ample allowance for intermediate consolidation, may be necessary.

Specifications
The specifications on the sand fill necessitated by the construction are generally related to the dimensional tolerances, the smoothness of the surface, the slope gradient, the density in connection with stability against failure and liquefaction, and the sensitivity to settlements. Integration of design, choice of material and construction method are of prime importance.

15.6 Building up above water level of sand fills constructed in water

In many cases it is necessary to build up a body of sand which is partly situated below and partly above water level. This can be the continuation of the underwater placement as mentioned in 15.5, but can also be the result of sand placement directly on an existing base situated some meters below water level. The most important difference with the placement as described in 15.5 is that for this situation another technique of sand placement has to be adopted; the underwater sand fill is incorporated in the construction above the water.

The abovewater fill will be built out with much flatter slopes than the underwater fill if an open fill area is applied. However, it can be guided into the desired trajectory by guide bunds made with bulldozers which move the sand into the steep bunds. When these guide bunds form a closed circuit around the fill area one speaks of a closed fill area. In that case the process water is discharged via an outlet. Through stacking of layer after layer of enclosed fill areas, the sand fill can be built up to large heights with steep slopes. Care must be taken that no pockets or zones of fines become entrapped in the fill area.

The possibility of re-shaping and compacting the sand from the water level upwards, using dry earth moving equipment, creates construction possibilities that are not possible in an underwater fill area. Namely steeper slope gradients, non-sensitivity to flow slides and minor settlement.

Functions of the sand

As before, the main functions are vertical support and load distribution. The drainage capacity (avoidance of excess hydrostatic pressure) also may play a role. Additionally the sand serves as a base for the often required slope protection. This has to prevent wave (ice erosion) at the water level, wind and rain surface erosion and in connection with the drainage capacity, compensate for excess hydrostatic pressure.

Side effects

Although the abovewater fill area (open or closed) is possible with many types of sand, especially for very fine sand and silty sand, the permeability may be a problem. The dewatering of the fill area is then very slow, preventing the dry earth moving equipment (bulldozers) from operating on the fill area. Re-shaping is then impossible. Moreover these types of sand generally cannot, or can hardly be compacted. In such a situation vertical drainage may offer a solution.

Specifications

In this case specifications are defined by the possibility of a need for mechanical treatment of the abovewater fill area. This especially concerns the extent of compaction. Furthermore the tolerances can be much smaller. Settlements and instability criteria are mainly determined by the underlying layers of sand and the subsoil. Grain size distributions and/or a maximum allowable silt content can be of importance.

CHAPTER 16

EQUIPMENT AND WORK METHODS FOR THE WINNING OF SAND

16.1 Introduction

For the hydraulic winning of sand various working methods can be adopted. To achieve high production rates suction dredging, cutter suction dredging and trailing suction dredging are the most important methods.

Apart from these methods, hydraulic and mechanical backhoes are employed for the winning of sand and gravel as well as grab dredgers and bucket dredgers. Backhoes are mainly used for the winning of sand in the dry and for winning in relatively shallow water, whereas grab cranes are employed for winning of sand and gravel on very large water depths. On the Dutch rivers also winning of sand takes place on a small scale using grab cranes in combination with inland barges.

The efficiency of the different methods depends on the circumstances in and around the borrow area. Here the following factors may play a role:

- composition of the sand in the winning area;
- depth and thickness of the dredgeable layers;
- location of the winning area with respect to waves and shipping;
- acceptable tolerances in the final structure;
- presence of unsuitable overburden;
- boundary of the future borrow pit.

Initially a general description is given of the sand winning methods followed by comments in detail on the above factors with respect to the most important sand winning methods.

16.2 General description of the most common sand winning methods

16.2.1 Suction dredging

With suction dredging, sand will be dredged by putting the suction tube deep (say 10 m) into the sand layer (see Fig. 69). Under the influence of gravity forces the sand departs from the slope (face formation) and flows downward in the direction of the suction mouth. In this way a so called „pit production" develops, which depends to a large extent on the total surface of the sand delivering slopes of the borrow pit. In case of well flowing sand and a sufficiently developed face this method provides high sand production rates. By moving the suction tube, which is often equipped with powerful jets, a

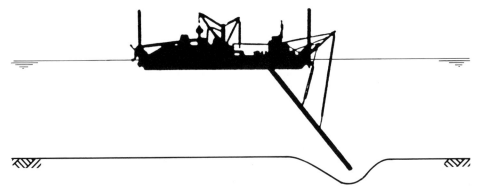

Fig. 69. Suction dredger.

certain pit production can be maintained. This production is preferably beyond the maximum production level of the dredger. On its turn the dredger's production capacity is limited again by the available suction- and delivery pump power.

16.2.2 *Cutter suction dredging*

With cutter suction dredging the suction tube is equipped with a rotating cutter head (see Fig. 70). The soil in front of the suction mouth is cut loose during the swing movements of the pontoon. The swing movement is initiated by means of the so called forward-side-winch-wires directly behind the cutter head. A spud-pole positioned on the ships aft functions as center of the swing movement.

With the cutter head the soil can be excavated layer after layer. With the winning of

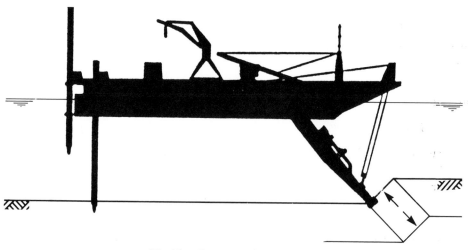

Fig. 70. Cutter suction dredger.

Cutter suction dredger.

sand using a cutter dredger, it is common practice to make use of the effect of face formation by applying a face height which is many times larger than the cutter head. Through this, higher production rates can be achieved.

By pushing the vessel forward against the spud pole after each sweep over a certain step length, the cutter head is brought forward. In this way the cutting production will be maintained. After a number of steps the spud pole will be hoisted, put forward and placed again on the bottom. During this procedure the secondary spud pole is placed on the bottom in order to fix the position of the dredger.

16.2.3 *Trailing suction dredging*

With trailing dredging of sand a draghead attached to a suction pipe is trailed over the bottom (see Fig. 71). Due to erosive forces at the narrow opening between the draghead and the bottom and/or the application of blades in the draghead a sand-water mixture is formed. This mixture is pumped in the hopper, in which the sand will settle and from which excess process water will flush overboard via an overflow. The use of an overflow provides also the possibility to improve the quality of the sand by spilling the fines over-board. Needless to say that this is also possible with a cutter or suction dredger when the sucked sand-water mixture is loaded into barges instead of pumped directly to the fill area via a pipeline. In this way the fine fraction stays behind as spillage in the winning area.

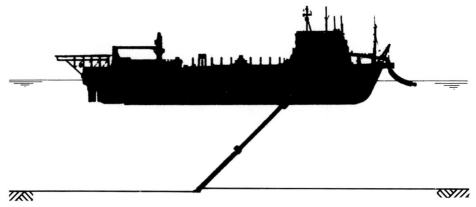

Fig. 71.   Trailing suction hopper dredger.

16.3   Composition of the sand in the winning area

Suction dredging
In case of sand with a low permeability and/or intermediate clay layers, face formation will be less well developed. In order to achieve an acceptable production level, a steeper and higher face slope has to be applied. Because of this a higher probability exists of the occurrence of uncontrolled large slides in the face, resulting in an irregular production process and possibly even damage to the suction pipe.
In case of intermediate clay layers these will be mixed with the sand during the face formation and pumping process and reduce the sand quality.

Cutter dredging
For less permeable sand the face formation will deliver a small contribution to the production. In this case the soil has to be „retrieved" with the cutter head. In less permeable sands large cutting forces develop during the cutting process and wear to the cutting teeth and blades occurs.

Trailing suction dredging
The trailing suction process is influenced to a large extent by the composition of the sand. With coarse sand the erosion process proceeds very well so that cutting is not necessary. With fine sand this is not the case and the use of teeth and/or blades may be necessary in order to maintain the suction process at an economic level. Also the settling process in the hopper worsens considerably with a decreased grain size of the sand.

16.4 Depth and thickness of the dredgeable layers

Suction dredging
With suction dredging a pontoon is used which is anchored on a wire system. With the application of a submerged pump on the suction pipe ladder it is possible to dredge sand from very large depths with an extended vertically suspended suction pipe. In this way depths can be achieved up to 70 m. However, for a sufficient production level a sufficient thickness of the dredgeable sand layer is necessary of at least approximately 10 m.

Cutter dredging
The thickness of the sand layer does not play a direct role in the cutting process at the cutter head, but is important for the maximum achievable production level as already indicated above. With the modern cutter suction dredgers sand layers can be dredged up to a depth of approximately 30 m. The minimum layer thickness for a reasonably economic dredging process using a cutter dredger is 2 to 3 m.

Trailing suction dredging
Typical for trailing suction dredging is the limited layer thickness which can be dredged per trail passage. Hence this method is extremely suitable for the winning of sand layers of a limited thickness at the sea bed. However, for an optimum production rate the winning area must be sufficiently large. In particular the length of the winning area must be large enough (minimal approximately 1500 m) in order to limit the loss of time due to manoeuvering as much as possible. The maximum suction depth of the largest modern trailing suction hopper dredger is approximately 50 m.

16.5 Location of the winning area in relation to waves and shipping

Suction dredging
From the anchored suction dredger the produced mixture can be delivered via a floating pipeline or by barges which are loaded alongside. The connection between the floating pipeline and the pontoon and possibly also the transition of a floating pipeline to a sinker pipeline on the bottom is sensitive to delays caused by waves (see also Chapter 14). The pitching movement of the pontoon can also cause damage to the suction pipe when this hits the bottom and is not equipped with swell compensators. However, the application of swell compensators on cutter dredgers and suction dredgers is still being developed and therefore only available by exception. For these reasons, in general, dredging operations have to be stopped for waves exceeding approximately 1 metre, whereas wave heights of approximately 0.5 m will stop the loading of barges.
The presence of a floating pipeline and anchor wires may cause hindrance to shipping. If the borrow pit is situated in a harbour area it may occur that the suction process is interrupted by passing ships. In this case it may be a problem to relocate the borrow pit. Also re-activation of the face may be a disadvantageous factor.

Cutter dredging
Here, the same considerations apply as for suction dredging. However, because of the swing movement in a sideways direction, hindrance to shipping in a narrow channel may be somewhat greater.

Trailing suction dredging
Because the trailing suction hopper dredger is equipped with a flexible swell compensated suction pipe suspension, the operational limit for the winning of sand is considerably higher. Wave heights of 3 m pose no problem to the suction of sand with the trailing dredger. In most cases however, the discharging process in shallow water or the coupling at sea on to a discharge pipeline may be the operational limit. Hence, these operations have to be halted when wave heights exceed 1 m.
The hindrance to shipping is minimal for a trailing suction hopper dredger as a result of the very good manoeuvrability of this type of vessel.

16.6 Acceptable dredging tolerances

Suction dredging
With suction dredging, differences in height of some metres may remain in the winning area. This has to be accounted for in judging the dredgeable sand quantities. Reduction of these height differences in the dredging cuts is possible by placing the cuts closer to each other however this results in loss of production. In these circumstances the use of a dustpan dredger may be considered. This has a very wide suction mouth (7 to 10 m) with which a reasonably smooth surface can be dredged.

Cutter dredging
The dredged profile with cutter dredging of sand is reasonably smooth because the cutter depth can be adjusted rather accurately and the irregularities between two swing movements are limited as a result of the shape of the cutter head (approx. 0.3 to 0.5 m). If it is a matter of face formation at which extra sand flows down the slope (cutter dredging in combination with suction dredging), a larger quantity of losses stays behind the cutter head.
The spillage layer which stays behind can be removed if necessary by finishing off the whole area again one time with the cutter dredger (a so called clean-up sweep). However, during this clean-up the production rate is considerably lower. Often in practice suction takes place at such a depth that the top of the spillage layer is still below the level which has to be delivered.

Trailing suction dredging
With trailing suction dredging it is not possible to produce an accurate profile. The reason for this is the passive following of the drag head of the sailing route of the hopper dredger which can be maintained with a limited accuracy of the order of 5 to 10 m. By

applying a well distributed pattern of sailing routes across the winning area, it is still possible to leave a reasonably smooth surface behind.

16.7 Presence of unsuitable overburden

Suction dredging
The presence of unsuitable overburden often means that this has to be removed before the winning of sand can start. For the removal of such a layer often a cutter dredger will be employed, whereby special attention will be paid to restrict the layer of dredge spillage. For thick top layers of clay, under special circumstances, the sand may be sucked from underneath the clay layer in distinct locations. In this situation the clay layer is temporarily supported through water injection. Because at each new location face formation in the borrow pit has to be started again the production level is limited. For this method also the dredgeable quantity is limited.
It is also feasible, in case of a thinner clay layer, to let this break off and to make use of the fact that the clay lumps „float" on the dense sand-water suspension within the borrow pit.

Cutter dredging
With cutter dredging in general any unsuitable top layer has to be removed first. The selective suction of sand from underneath a clay layer is not possible because the clay layer prohibits the sweep movement.

Trailing suction dredging
With a trailing suction hopper dredger it is in principle possible to remove the unsuitable top layer first. However the top layer of the dredgeable sand will be lost because this is mixed with the unsuitable soil.

16.8 Boundary of the future borrow pit

Suction dredging
The slope gradient which results after face formation has finished, depends mainly on the applied face height, the grain size diameter of the sand and the packing density. In some instances even face slopes of 1 to 20 have been measured (Venserpolder [53]), along which the sand-water mixture produced by the face flowed down towards the suction mouth. The active sand producing face slope is always steeper than approximately 30°.
If the slopes of the borrow pit have to remain within prescribed boundaries then the face height along these boundaries has to be limited. This results in lower sand production rates.

Cutter dredging

For cutter dredging with a high face in principle the same considerations apply as for suction dredging. However with a cutter dredger it is possible to cut a stable side slope according to the following two procedures:

- by hoisting the cutter head after each sweep along the slope;
- by making a step profile in the slope, whereby the ending of the cut layer must be such that the average slope is obtained with a minimal disturbance of the subsoil (the so called „boxcut", see Fig. 72).

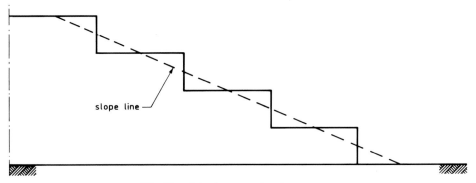

Fig. 72.   Principle sketch of a boxcut.

Trailing suction dredging

The negative phenomena for the formation of slopes mentioned above, hardly apply for trail dredging because the layer thickness removed each time is very limited.

CHAPTER 17

WORK METHODS FOR SAND FILLING

17.1 General

Various methods for sand placement can be distinguished which can often be performed with different types of equipment. For instance sand can be bottom discharged with various types of barges equipped with various types of valves. In addition methods using a grab crane may be employed and methods via discharge pipes which may or may not be suspended under water. The most common methods are discussed in this Chapter.

The accuracy of location control, the workability and the draught differ for each piece of equipment. These aspects are explained by type of equipment in 17.2, 17.3 and 17.4 respectively. Also the berm height up to which sand can be discharged and the control of the discharging differ for each method. These aspects are explained in 17.4, 17.5 and in Table 6.

17.2 Positioning in place of the equipment

Bringing the vessel into position as accurately as possible is of great importance for the placing of sand in a fill area, because this will result in large savings on the required sand volume.

Bringing into position means manoeuvering the piece of equipment by which the sand is placed into the right position and maintaining this position. For discharging while sailing this means keeping the discharge vessel on its right course.

For the various types of equipment a summary of the positioning parameters is given in Table 4. These parameters depend on the weather, experience of the crew etc., so that they cannot be regarded to be generally applicable.

For the sake of completeness it should probably be stated that the mentioned position accuracies apply to the relevant equipment and not to the discharged sand. The determination of position and the surveying of this position in the field are idealised in Table 4. Hence, inaccuracies in the determination of the position have to be superimposed on these values.

The accuracy of positioning depends also on the accuracy of the positioning system. Table 5 gives a summary of the most common systems. The presented accuracies are based on an optimal configuration of the shore stations.

Table 4. Position accuracy for discharge equipment.

| equipemt | position accuracy for* | | wave height | saling along straigth line possible | draught laden |
	$H_s=0.3$ m (m)	$H_{s\,all}$ (m)	$H_{s\,all}$ (m)		(m)
hopper dredger + dynamic position	5	10	1.5	yes	4–9
hopper dredger + bow thruster	15	25	1.5	yes	4–9
barge, self propelled	15	30	1.0	yes	3–5
barge on wires	1	1	0.5	yes	3–5
barge on spud	0.5	0.5	0.3	no	3–5
barge with tug	20	30	0.7	yes	3–5
pontoon on wires	1	1	0.5	yes	1–2

* position accuracy: the mentioned distances are measured for instance from a centre line
H_s: significant wave height
$H_{s\,all}$: maximum allowable significant wave height
** barge: barge with hopper and not self propelled
pontoon: barge without hopper (hence only auxiliary equipment)

Floating pipeline subject to wave action.

The systems mentioned in Table 5 in the group medium and long range, are all based on the measurement of phase differences with the exception of the GPS system. The mentioned micro-wave systems are based on the measurement of the time required for the transmission of an acoustic signal to a fixed station and the retransmission of that signal. Range bearing systems are based on the measurement of an angle with respect to a fixed orientation and a distance.

Table 5. Positioning systems and accuracies.

system		reach	accuracy	remarks
medium and long range system	GPS (global position system)	worldwide	in maritime applications: appr. 100 m, in the years 90 can develop to very accurate	US Navy System, in 1993 worldwide
	Omega	international	appr. 100 m	US Navy System
	Loran C	international	appr. 100 m	US Navy System
	Decca	international	appr. 30–100 m	accuracy dependant on location
	Pulse 8	appr. 400 km	appr. 30 m	
	Hyper-Fix	appr. 300 km	appr. 5–10 m	partly North Sea
	Syledis	100–250 km	appr. 5 m	Syledis chainage across large part of North Sea and Wadden Sea
micro-wave system	Trisponder 540/217	appr. 80 km	appr. 1 m	
	Motorola Miniranger	appr. 35 km	appr. 3 m	
	Micro-Fix	appr. 60 km	appr. 1–2 m	
range bearing system	Artemis	appr. 30 km	0.5 m/km	
	Polar-Fix	appr. 5 km	0.5 m/km	
visual system	sextant	max. 1 km	optimal 5 m	
	flagpole + distancemeter	max. 1 km	optimal 5 m	
	transit makers	max. 1 km	optimal 5 m	
	laser + distancemeter	3–5 km	<0.5 m	

17.3 Workability

See for the workability also Chapter 14.

Waves

Table 4 indicates the level of significant wave height $H_{s\,all}$ that can be worked reasonably well. The definition is based on the supposition that the direction and the period of the waves are most unfavourable from the equipment point of view. A problem that occurs is the relative movement of the different equipment pieces (for instance interaction between a pontoon and the discharge pipe).

Current

Currents can also influence the workability. Forces in wires and winches as a result of the current against the pontoon can be unacceptably high. It can also be the case that the current forces on floating pipelines which may be present become too large. Stronger wires, stronger winches and stronger anchors may be the solution. Currents can also influence the positioning accuracy. For current velocities larger than 2 m/s dynamic forces can occur on the pipeline causing the pipeline to jump.

Temperature

In extreme circumstances the temperature also can be a limiting factor, especially because of ice formation on the equipment.

Wind

In some cases the work may be hindered by strong winds.

17.4   Draught and maximum berm height

The draught (fully loaded) of the various pieces of equipment is mentioned in Table 4. However, the difference in draught between a fully laden and empty vessel can be very large.

The maximum berm height, this is the berm height which can be obtained by building up using a certain filling method, depends on the draught of the vessel. Also the wave height and the possible valves protruding from under the bottom plate of the vessel determine the height of the berm.

17.5   Control of the discharge process and bridge formation

In Table 6 the controllability is indicated for various types of discharge floating equipment, the probability of bridge formation and the obtainable berm height. To illustrate this Table reference is made to Figure 73.

Controllability

The controllability of the discharge process is defined as the capacity to discharge the theoretically required volume of sand at a flow rate determined beforehand (m³ sand per s). This is an important factor if an accurate sand fill has to be built up.
A proper control can be spoken of when the sand can be discharged in a very controlled manner.

Table 6. Controllability, bridge formation and berm height for various discharge vessels and valves.

method	common name	sketch (Fig. 73)	application			control-lability	danger of bridge forma-tion or bad discharging for fine sand	maximum berm height under water level, whereby draught = draught fully laden (m)
			hopper	barge	pontoon			
1	bottom sliding door	a	+	+		moderate to bad	yes	draught + 1.0 + H_s
2	bottom doors (dumping barge or dumping barge with recessed doors)	b, c	+	+		bad	yes	draught + valve length + 1.0 + H_s
3	cone valves	d	+			moderate	yes	draught + valves + 1.0 + H_s
4	split dumping barge	e	+	+		moderate to bad	yes	draught + 0 + H_s
5	grab crane	f				n.a.	no	n.a.
6	stone dumping barge	g		+		very good	n.a.	0
7	spraying/rainbowing	h	+		+	n.a.	n.a.	n.a.
8	pipe under water	i	+		+	good	no	draught (0.5) + H_s
9	pipe under water + diffusor	j			+	very good	no	draught (0.5) + H_s
10	pipe above water	k				moderate	no	n.a.
11	grab crane + pipe	l				n.a.	no	n.a.

149

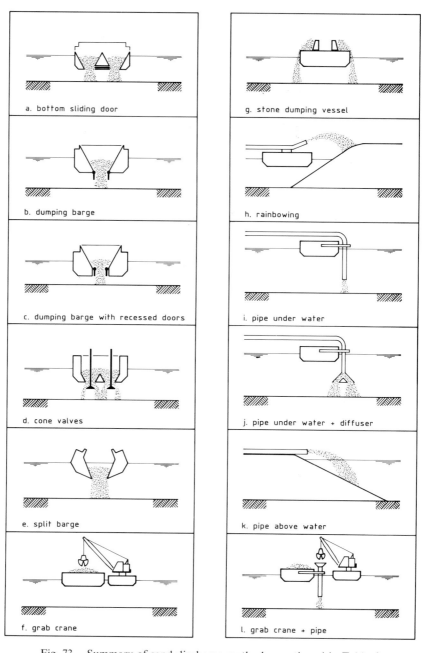

Fig. 73.   Summary of sand discharge methods mentioned in Table 6.

According to work method the controllability can differ strongly. A „stone dumping vessel" generally is very well controllable, a barge with bottom doors however is in general very badly controllable. This controllability not only depends on the type of equipment, but also on the type of soil.

If the system is badly controllable, the possibility exists that the discharging is too irregular and also often takes place with too large a flow rate, which means that crater formation occurs.

Bridge formation

Bridge formation or arching can occur during the unloading of barges or hoppers. Bridge formation is the building up of forces in the material to be bottom discharged, as a result of which, after a small quantity has been discharged, the remainder of the material stays behind as a bridge. This phenomena can occur with gravel and coarse sand and for sand with possibly some clay content which has a low permeability. This problem can be prevented by using jets in the hopper which destroys the bridge.

Other factors

The accuracy of the profile which must be produced in the end depends on the above mentioned factors. Additional factors also include the chosen work method, the shape, surface and volume of the fill structure to be made as well as the experience of the crew and the site manager.

CHAPTER 18

QUANTITY MEASUREMENT FOR CONSTRUCTION
AND PAYMENT

18.1 Purpose and location of the measurement

The measurement of the quantities of handled soil is important both for the payment and the construction of a project. For the construction also insight in the production capability of the dredging equipment is of importance. In principle the measurement of the dredged quantities of sand can take place at three points in the production process:

a. in the winning area by in and out survey of the borrow pit;
b. during the transport:
  – in the discharge pipeline by measurement of the mixture velocity and the density;
  – in the hopper dredge or barge with soundings;
c. in the fill area by in and out survey.

The quantities as mentioned under a, b and c will generally differ as a result of the losses, bulking etc.(see 18.2).

18.2 Inaccuracies in the measurements

Apart from certain inaccuracies of the registration instruments themselves, the following reasons for deviations can be mentioned:

For measurements in the borrow pit:

– deviations on the information on tidal levels because of hydraulic gradients and swell;
– deviations between soundings by hand or echo-sounder (see 18.3);
– sedimentation and erosion in a river or tidal area;
– back-flowing of sand in the borrow pit.

For measurements during transport:

– in barges: bulking of soil during transport in the hopper dredge or barge (see 18.3);
– in the discharge pipe: lagging behind of the coarser fraction on the bottom of the pipeline compared with the average mixture velocity in the pipeline (transport factor).

For measurements on the fill area:

– consolidation of the sub-soil;
– sand losses via process water;

- sand losses by wind erosion on the abovewater fill area;
- erosion or sedimentation during sand placement in flowing water;
- losses by currents or waves.

The selection of the measuring system depends on the circumstances of the works and must be such that the deviations and inaccuracies are minimal.

18.3 Bulking of the soil

Due to the change of the soil conditions during excavation, transport and sand placement, so called bulking occurs. A change of the soil volume occurs as a result of the in or outflow of water or air in the pores. With sand the volume may increase with a factor up to 1.2, especially for de-mixing of well graded sand. However, the final bulking depends on the original packing condition, the applied work method and compaction method. For the winning of loose sand with a trailing hopper dredge a factor of 0.8 may occur (see also Appendix A).

18.4 In and out survey by means of soundings

For vertical sounding a sounding lead or echo-sounder may be used. On this subject the following remarks can be made:

sounding lead:

- slow and labour intensive, can only be applied for small borrow pits and limited depths;
- because of current influence sounding deviation up to approximately 0.2 m;
- sounding lead may roll down a slope;
- when approaching a slope the lead lags behind in the deeper part;
- at greater depth the bottom is difficult to sense.

echo-sounding:

- provides a continuous signal;
- result depends on the sound velocity in water which depends on the salt content and temperature;
- beam width and soil penetration depends on the frequency;
- slopes are sounded flatter and more shallow;
- false echos on light mud;
- with irregular bottoms always the highest spots are sounded.

During the execution of a project the echo-sounder is calibrated regularly with a so called „barcheck". A steel girder is lowered in the water with the depths measured with a tape attached to the girder. The bar is then sounded with an echo-sounder. With this method, apart from the zero fixation, the influence of the salt and temperature on the sounding can be compensated as well.

For the in and out survey the horizontal positioning is of importance, especially with the sounding of the slopes. Apart from the radio positioning system, with accuracies varying from some metres to less than 0.10 m, optical systems exist such as the sextant with accuracies of some metres.

18.5 Measurement in the hopper or barge

The measurement in the hopper or barge can be related to loaded cubic metres or tonnes. With sounding in cubic metres, the sand level is determined using a sounding rod or sounding lead and calculated to volume in m^3, using a sounding table. The obtainable accuracy is of the order of magnitude of some percent.

For payment of barge loads in tonnes, this load can be determined by reading the vessels draught on its watermarks before and after loading. However the density of the surrounding water must be known for this and the reading should also be done in quiet water. A disadvantage of this visual measuring system is the limited relative accuracy with which the difference in draught is determined.

For the large modern trailing suction hopper dredgers, nowadays the draught can be determined reasonably accurately by means of hydrostatic pressure gauges at the bottom of the vessel. Together with the measurement of the water level inside the hopper and a known value of the water and sand density, the weight in tonnes of loaded solids can be calculated. However, the difficulty of translating these tonnes into m^3 in situ material remains. As it happens, the result of this calculation depends strongly on the input value of the in situ density.

18.6 Measurement in the discharge pipeline

The mixture velocity in the pipeline can be measured rather accurately with an electro-magnetic flow meter. The density of the mixture can be measured with a radio active density meter on the basis of absorption of gamma radiation by the sand-water mixture. In practice it appears that silt-sand mixtures, clay balls and gravel stones etc. may cause disturbances.

18.7 Measurement on the fill area

When the fill area is raised above water, the survey may take place using water level instruments. With consolidation of the subsoil, settlement beacons have to be placed at regular distances on the original ground level. Herewith, the average level of the bottom side of the fill area can be determined. The accuracy of the result is approximately 0.1 m.

CHAPTER 19

QUALITY ASSURANCE AND PROCESS CONTROL

19.1 Introduction

For projects it is becoming more and more common practice that designers and/or contractors are requested to prove that the quality of the result is assured and verified. There are a great number of factors which influence the quality of a project, for example internal factors such as personnel, expertise and equipment, but also external factors which can vary considerably. In general, the quality of a product cannot be achieved just by an inspection of the final product; interim control and adjustments are often necessary. Quality control is used for this.

A brief description of quality assurance is:

quality assurance = process control

For the quality assurance of a project a quality plan has to be formulated in which the complete quality control process is laid down. Through regular inspections, during which verifications are made according to a checklist formulated beforehand, the process can be controlled. In literature this is known as the so called quality control loop (QC-loop), this implies: action-verification-improvement-action, etc. (see Appendix D).

The strongly varying external factors lead to uncertainties. By using reliability analyses activities subject to risk can be traced and if necessary improved.

In Appendix D the definition of quality assurance is discussed in general. This Chapter discusses how quality assurance can be applied in practice.

19.2 Quality assurance in practice

As mentioned before: quality assurance = process control. Assurance of the quality of certain work begins at the start of the project and has to be maintained right through the project. The following phases can be distinguished:

phase 1. formulating needs;
phase 2. design;
phase 3. work preparation;
phase 4. actual execution;
phase 5. commissioning and verification;
phase 6. follow up and maintenance.

Often phases 1 and 2 are indicated as design phase, phases 3, 4 and 5 as construction phase and phase 6 as the maintenance phase. The different phases are often executed by

155

different parties. In phase 1 the client formulates his needs. The design may be carried out by the client himself, by a consulting engineer as well as by one or more contractors; also the work may be executed by one or more contractors. Traditionally design and specifications (the technical and administrative conditions of the project) are made by the client or an advisor. After that the contractor carries out the work according to the specifications and the client verifies whether the execution and the product are in accordance to the specifications.

Nowadays it has become common practice, that the client only formulates his needs and that another party (or a combination of parties) looks after the design as well the construction. These parties have to prove (beforehand!) that the work will satisfy the formulated requirements by showing an available, implemented and efficient quality system. Hence, the responsibility for the quality remains with the designers and supervisors of the work. The client must clearly formulate his requirements and provide sufficient data so that the worksmanager responsible for the execution also can really deliver the required quality. In his selection procedure the client will not only consider the price but also that the required quality is met and assured. An important criterium for this is the quality plan of the parties involved. This plan must clearly indicate the organization, the responsibilities, the methods and the reporting system.

It must be indicated how the quality of products is controlled within the contractor's organization but, more important for the client, he must be able to show that the work meets the formulated requirements. To this end the reporting system is an important part of the process because participants must be able to check the work during the different phases of the project and if necessary improve the quality.

In the next section the application of quality assurance is described for the relevant processes of each of the already mentioned phases of constructions where sand is placed in or under water.

19.3 Formulating needs

During this stage the needs have to be defined and an analysis of the functional requirements has to be made. In Chapter 15 a summary is given of the different construction types, the most important functions and the corresponding requirements. It was also indicated which side effects can occur. According to the quality control loop (see Appendix D), the next step should be whether all the items of concern are identified and described and whether the expected side effects are acceptable. The requirements must be clearly formulated so that in the next phase designers and other parties can work from this.

19.4 Design

19.4.1 Design process

The total design process must be included in the quality plan in which the specific quality control is defined within the frame work of the quality system. The plan

describes the process how the various design aspects are dealt with, how quality verifications and audits are organized and which tolerances are acceptable. For control of the total process, the QC-loop applies to each step. Not merely an extensive verification when the design is finished, but a continuous (documented) verification of the quality level on the requirements and the basic assumptions should take place.
For each project the design process consists of the following steps:

- analysis of functional requirements;
- collection of data;
- definition of boundary conditions;
- development of alternative designs;
- rough calculations of the various alternatives;
- cost estimate;
- verification with respect to feasibility and workability;
- reliability analysis;
- verification of alternatives;
- final selection;
- design calculations, determination of dimensions and work methods, reliability analyses;
- detailed cost estimate;
- contract documents, specifications and drawings.

The first 10 steps until final selection can be regarded as an iterative process.
Regarding the design of a sand structure in or below water, attention has especially to be paid to:

- choice of sand winning area and sand winning method;
- choice of transport system (hopper/pipeline);
- choice of placement method.

In the following the different parts of the design process is discussed, the possible verifications and the assessment of acceptable tolerances.

19.4.2 *Functional requirements and data*

The design process starts with an analysis of the functional requirements, as these result from the formulated needs. In order to be able to quantify the functional requirements data are needed.
For the collection of data a soil investigation is very important. In Chapter 3 the necessity of a soil investigation in the borrow area is mentioned and methods of such an investigation are discussed. A soil investigation is also required in the discharge area, for example to find out the danger of flow slides. In Chapter 14 the relevant boundary conditions for the execution of a project are mentioned; for this also data are required.
For the collection of data it is important that during the design process a distinct plan

can be formulated so that uncertain and unknown aspects of the design are known. In this manner an evaluation of the project can be made after the execution and thus the design and the execution method of a next project may be improved.

19.4.3 *Development of alternative designs and considerations*

For the development of alternative designs the execution method is of great importance. The requirements and the boundary conditions may directly lead to the selection of a certain execution method. Whether a project is executed in sheltered or in exposed surroundings can be decisive for the type of equipment, the most favourable placement method and the workability (see for example 21.3 and 21.4). The design alternatives will be composed of different items, resulting from the different processes which are relevant for the project. The most important are: choice of the sand winning area and winning method, choice of the transport and placement method.

The main purpose of all the calculations of alternative designs is to obtain a first impression of quantities and dimensions. According to the quality control loop each individual alternative design has to be reviewed and verified concerning its functional requirements and formulated needs. At this stage it is good to recognize the importance of working according to a well implemented quality assurance system: by reviewing systematically the alternatives one by one the risk is minimized that the chosen alternative is based upon the wrong assumptions. This means a possible cost saving on the extensive calculations for the final design. A reliability analyses may be used for a proper consideration of the alternatives.

19.4.4 *Final design*

For detailed calculations the theory as described in the Chapters 2 up to 12 can be used. A summary of this theory is presented in Chapter 13 including a method for calculating the various parameters. By means of a reliability analyses which parameters are most important can be checked.

The result of the design process must be laid down in the specifications of the works. For the quality assurance during the execution it is important that the design clearly indicates what has to be constructed or delivered, how this must be realized and what requirements the project has to satisfy. Essential for a sound design is the use of a good documentation system for the following reasons:

- every one involved must be able to easily find relevant information;
- the manner in which the (partial) designs are realized must be traceable;
- during change of personnel the tasks must be smoothly transferable;
- during the execution effective use of the already gathered knowledge has to be made;
- the loss of papers and other data carriers by fire or any other disaster has to be prevented;
- design documentation has to be transferable to all concerned.

19.4.5 *Verifications*

During the whole design process the following verifications can be carried out:

- self-check, for the support of which the designer can make use of the guidance of team members and of internal project meetings or the consultation of specialists possibly with the aid of check-lists;
- verification of calculations by third parties, for which distinction can be made in different levels of verification, depending on the relevance (risk) of the (partial) design;
- design audits: during a design audit a (partial) design will be judged by internal and/ or external specialists; it often concerns the checking in broad outlines of a (partial) design; results of the design audits are recorded and included in the documentation system.

19.4.6 *Tolerances*

An important aspect of the quality is the acceptable tolerance. It has to be observed carefully for which items tolerances have to be defined. Tolerances are important both for the design and the construction stage. The ideal situation is obtained when both phases are smoothly linked with each other and a continuous process results of: exchange of experience, adjustment and optimization. The best set of tolerances emerges when continuous consultation takes place between designer, client and contractor.

Recommendations for tolerances are:

- Make an inventory of items for which tolerances must be defined.
- Quantification of tolerances.
- Verify tolerances consciously on their feasibility. Too strict as well as too lenient tolerances will result in loss of quality (often tolerances are unnecessarily strict based on the idea that „then something good will be obtained").
- If possible, make use of norms, certificates, practical guidelines and the like for the assessment of tolerances. Indicate clearly where stricter requirements are put forward. Make an inventory of the regularly returning tolerances with a verification of the feasibility and formulate these on tolerance sheets.
- Identify critical tolerances and indicate how often these are verified.
- Indicate in the contractual documents or specifications how deviations have to be dealt with (if the defined tolerance cannot be achieved).

For the formulation of tolerances the following has to be taken into account:

- requirements of the client;
- production possibilitie;
- cost consequences.

It is the task of the designer to check whether certain prescribed tolerances are relevant for the specific situation. If these tolerances are not verified and blindly copied, problems in the construction phase will result.

19.5 Work preparation

As stated already: quality assurance is process control. Therefore during the work preparation the execution process must be analyzed and it must be formulated how this process has to be verified.

The work preparation consists of the following steps:

– formulation of project manuals, in which the work methods and the planning are described in detail;
– reliability analyses;
– formulation of procedures which indicate how the quality of the work can be verified;
– formulation of safety procedures;
– preparation of personnel, equipment and materials on the actual execution.

During the work preparation reliability analyses of the work method are important. By investigating which elements of the work method entail certain risks, an impression of preventive and/or responsive measures can be obtained which might be necessary during the execution.

In the quality plan the working methods are bundled. This plan can only be formulated when the working methods are completely known. The quality plan must also indicate how the execution is controlled and verified. This part of the quality plan must be formulated during the work preparation and must be complied with during the execution.

In the procedures the subsequent quality activities are described. The quality parameters of the execution activities, such as winning, transport, placing etc. must be defined for this. These are parameters which influence the quality of the process (for instance during sand winning the grain size distribution of the sand). Thereafter it is decided how and to what extent these parameters can be controlled respectively verified. Which of the above mentioned parameters must be measured and how, is indicated in test procedures. These procedures must also provide a description of the recording methods. The recording of process parameters and quantity measurements needed for the execution and payment, is dealt with in the Chapters 11 and 18.

In Figure 74 an example is given for the setting up of quality plans for a certain part of a large project for which the design as well as the execution is taken care of by one party. For this work a project quality plan was formulated for the different project items. Within this project quality plan execution and test procedures are formulated for the different items of that part of the project. To apply the above terminology: in this example the working methods are described in the execution plans and the procedures related to tests and measuring systems are described in the test procedures.

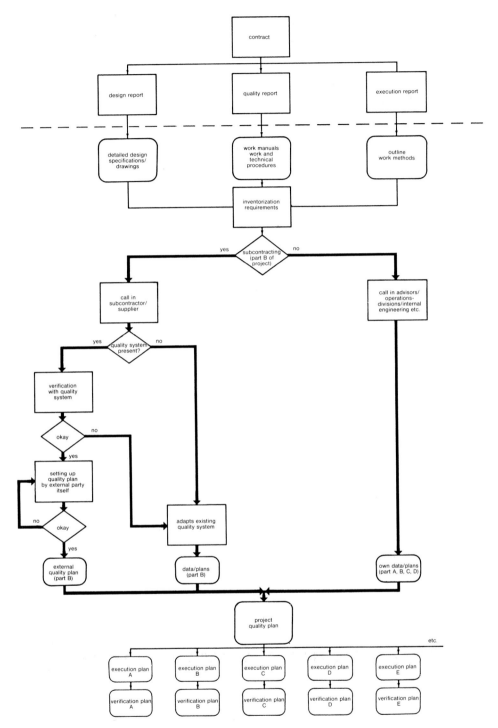

Fig. 74. Quality plans.

The contents of the project quality plan in this example include:

- introduction;
- organization and tasks;
- requirements, boundary conditions and permits;
- size of the project;
- planning of related execution plans;
- requirements with respect to safety, labour circumstances and environment;
- list of applicable documents;
- list of applicable procedures;
- list of subcontractors/suppliers.

and the contents of the execution plan in this example include:

- introduction;
- requirements and boundary conditions;
- execution;
- tolerances and measures when exceeded;
- items yet to be formulated.

A process verification plan

- is supplemented to a project quality plan or execution plan;
- sorts out and describes tests;
- describes the tests to be executed, time and location of the tests and surveys;
- defines responsibilities, test criteria and test procedures;
- serves as a source for a birth register (as part of the as built documentation) of the project;
- serves as a legal document after filling in of the test forms and signing of the testing officer and the person in charge;
- is standardized because of the use of standard test forms;
- indicates mile stones and pertaining actions;
- provides specific tolerance requirements.

This is only an example, for each project the quality plan is defined in a different way. However it must always consist of the items mentioned in the norms and must clearly indicate working methods and procedures.

19.6   Construction

Whether the quality of the work is well controlled can at best be observed during the construction. However one has to realize that success in this phase depends for a great deal on the quality of the preceding phases. By paying sufficient attention from the beginning to quality control and aiming at a consequent implementation of quality verification, (unavoidable) corrective measures will be limited to a minimum. This is especially important in the light of the most crucial factor during the execution: TIME. As the work progresses the factor time becomes more and more important. A lot of money can be saved by avoiding mistakes and will in this phase promote the quality of the work.

The complete execution process must be described in a method statement as part of the work preparation. In the quality plan it must be defined WHAT, HOW and HOW OFTEN verifications have to be made.

During the execution all procedures and working methods, as described in the quality plan, must be executed. In order to verify the process, the process parameters will have to be measured and it must be known how to act in case of deviations. Because the process consists of different items also different verifications have to be carried out. This is illustrated in Figure 75, which shows a summary of what items among others have to be verified. This list is not complete, moreover the quality plan must also

1a. Equipment (vessels,
   auxiliary equipment,
   pontoons, pipelines etc.):
- right equipment
- technical condition
- safety

1b. Subcontractors:
- for instance reliability

2. Personnel:
- organogram
- right persons
   (for the duration of the work)

3. Borrow area:
- equipment on right location
- maximum suction depth
- slopes
- soil investigation, verification
   silt content
- stability adjacent structures
- maximum volume of the borrow area
- excess noise
- execution in accordance with design

4. Transport:

4.1 Hydraulic aspects:
- leaking pipelines
- pre-compression of pipelines
- anchorage of floating pipeline
   (current, waves)
- spare pipeline present?

4.2 Water based transport:
- improvement of grain size
   distribution by overflowing of barges
- non leaking barges (sand/water)
- volume measurements
- sailing speeds
- manouevrability
- draughts
- de-mixing in barge or intermediate
   fill

4.3 Intermediate fill:
- location
- possible change in quality
   (silt content)

5. Sand placement:
- anchor management
- discharge velocity
- measurement of profiles/holes
- density/cpt's
- tolerances/smoothness
- grain size
- sand outside profile
- silt contents/enclosures
- slopes
- turbidity of surroundings
- contamination
- hydrostatic pressure
- consolidation measurements

6. Planning: Is the superintendent
   following the planning?

Fig. 75. Exemples of items which have to be verified with quality assurance.

provide descriptions as to how verifications are carried out and how often. In the Chapters 11 and 18 it is described how various measurements can be performed.

Apart from requirements concerning the functions of a construction, also requirements with respect to the side-effects may exist. These may occur in the construction stage as well as in the final product. These side-effects are dealt with in Chapter 15 for each construction type. The corresponding requirements may lead to extra verifications of the regularity with which the sand is placed, verifications of the sensitivity to consolidation such as by monitoring hydrostatic pressures etc.

During the work preparation detailed working methods are formulated, which include the expected durations of the different phases. These durations are determined on the basis of predicted sand production rates and losses. For this mathematical models are used with the process parameters as input. It is important to compare the progress of the project with the predictions and in the event of deviations to check by what these are caused. Only when the cause of the delay is known, it can be predicted more accurately whether this tendency will be there for the rest of the project. Depending on the importance of an accurate planning it is therefore recommended to determine the descriptive parameters during the progress of the execution of the works and with that verify the predictions of this progress and adjust these if necessary. This may also lead to adjustments in the work methods in order still to be able to realize the planned completion date. Moreover for a proper determination of the parameters, calibration of the applied prediction models may take place either as evaluation afterwards or, for a large project, during the execution.

19.7 Commission and verification

Commission can only take place when the project meets the required quality. All activities, reports and verifications must be completed.

Verification is an aspect of quality which is often neglected but is an essential part of the total quality control. The verification procedure should comprise the following:

- verification of the measurable product characteristics, as described in the quality plan of the project;
- analysis and verification of the measured process parameters with respect to the quality of the work;
- analysis and documentation of the method of execution.

The result of the verification procedure must be documented in a product verification report, the so-called „as built report". This report requires the approval of both the project leader and the client. Recommendations of possible improvements in the future should be mentioned in this report. In this way all collected data and gained experiences can be used to up-date the existing knowledge and improve on future designs.

CHAPTER 20

CASE HISTORIES

20.1 Drilling island Issungnak, Beaufort Sea, 1979

20.1.1 *Introduction*

Expectations of promising oil and gas reserves have sparked off an extensive programme of exploration drilling along the coasts of Alaska and NW-Canada. For that
purpose various drilling islands were built (and removed again) during the period 1976
to 1984. Due to the unique external conditions in this polar area new types of structures
had to be designed for which the following particular conditions had to be formulated:

- withstand ice forces caused by ice fields and ice bergs during the polar winter;
- withstand wind, current and wave attack during the summer;
- construction within a very short period (open water conditions);
- removal relatively simple;
- construction with locally available material (transport to the area is exceptionally
  difficult).

20.1.2 *External conditions*

During winter polar ice drifts into the coastal waters. During summer the ice retreats
resulting in an approximately 100 miles wide strip of open water. This applies to the
period between mid June and the beginning of October. During on shore winds the
polar ice drifts from the pack ice towards the shore. Supply of material over sea is only
possible if the sea lane around Alaska is ice free. This is sometimes only the case during
a few weeks per year.

The average daily temperature in January is approximately $-29\,°\text{C}$ and in July $+14\,°\text{C}$.
During summer the prevailing wind direction is east/northeast averaging 3.3 m/s. The
temperature of the water varies between $-1.5\,°\text{C}$ and $+1\,°\text{C}$. Most of the time the current
velocities are small and can occasionally be up to 1 m/s. In general the sea is calm, but
sudden squalls can occur causing the wave heights to rise beyond 1.5 m within the hour,
at wind velocities exceeding 30 knots.

The planned drilling island Issungnak is situated approximately 80 km northwest of
Tuktoyaktuk harbour in a water depth of 19 m. The tidal amplitude is within the order of
0.12 m.

20.1.3 *Structure to be made*

The planned structure consisted of a circular island with a diameter of approximately
100 m at 5 m above sea level (in a later stage it was decided to raise the island to 7 m
above sea level). In the wave and ice zone an erosion protection had to be applied. The
sand could be retrieved from underneath a top layer (overburden) with a thickness

165

of maximum 10 m. A suction depth up to 50 m could be applied for sands of 200 to 250 μm.

20.1.4   *Types of equipment and work methods*

For the sand placement for this type of island, experience was obtained during preceding years with the construction of the islands Arnak, Kannerk and Isserk in water depths of 9 m and 13 m. In the meantime it had also appeared that sand winning could be hampered by the presence of permafrost. For the previous constructions the same equipment was available i.e. split barges for bottom discharging as well as a stationary suction dredger (Beaver Mackenzie) suitable for winning at great depths, for pumping directly to shore and for loading barges.

Where required this equipment was re-enforced to withstand ice pressures and made resistant to the extreme low temperatures.

Because the island was constructed in greater water depths than previously, much more sand was required. Furthermore the sea conditions were far worse (especially during gales) causing the island to be attacked by winter ice from land and by dynamic transition ice.

20.1.5   *Construction aspects*

In order to limit the sand volume it was important that a minimum quantity of sand was deposited outside the required profile. In this respect it appeared that sand placement was critical for two reasons.

Firstly because of the required large sand volume (approximately 4 million m^3). Due to the short workable period, the construction had to be completed during two summer seasons. During the intermediate long overwintering period the constructed sand profile could erode.

Secondly it appeared that especially during the phase of raising the construction through the water level, a large increase of the plateau occurred. During gales wave heights up to 5 m occurred, washing away approximately 25,000 m^3 of sand daily from the top of the island.

From mapped soundings it appeared clearly that the massive quantities of sand required were owed especially to the natural developed beach profile (see Fig. 76). The underwater slopes realized with this construction method, varied between 1:10 and 1:18 with occasional slumping to 1:25 at the toe. A typical cross-section is given in Figure 77. All sand below 10 m water level was bottom discharged.

20.1.6   *Feedback with theory*

For feedback with theory, reference is made to 20.2 where the construction of the island Uviluk will be dealt with (see 20.2.6).

166

Fig. 76.   Three dimensional illustration of drilling island Issungnak.

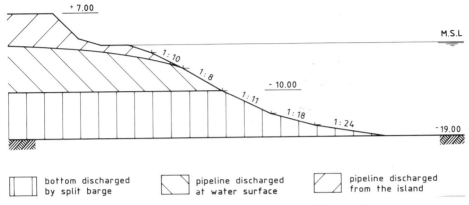

| bottom discharged by split barge | pipeline discharged at water surface | pipeline discharged from the island |

Fig. 77.   Building up of the artificial island Issungnak.

20.2   Drilling island Uviluk, Beaufort Sea, 1982

20.2.1   *Introduction*

The desired drilling location is situated approximately 75 km ENE of Issungnak (see case history 20.1) under comparable circumstances. However the local water depth is greater namely 31 m. Because of the difference in water depth of 12 m, sand quantities would increase enormously if a design identical to Issungnak would have been adopted. Here a different construction method was applied aimed to avoid in the first place beach formation at the water surface and in the second place to realize the steepest possible slope.

The first mentioned objective was achieved by finishing the sand berm at 9 m below sea level followed by placing a caisson on the submerged berm. As a result of this design a vertical surface at the water level passage is made which gives a considerable increase of the wave load on the structure. With respect to ice loads it should be stated that on the one hand the exposed surface of the obstruction is smaller but on the other hand a large shearing force on the caisson occurs. This again, reduced the piling up of drifting ice previously observed on the beach and the flat fore shore.

The second objective was achieved by placement of sand in a different manner (see 20.2.4).

167

20.2.2 External conditions

With respect to the external conditions practically the same conditions apply as refered to in case history 20.1. However, owing to a slightly more northerly position in deeper water, the chance of unworkable weather was somewhat greater.

20.2.3 Structure to be made

The structure consisted of the following items:

- sand berm to be built up to 10 m below water level;
- construction of this submerged berm, with slopes of 1 : 5;
- creation of a flat finishing of this submerged berm for a proper support of the caisson with dimensions of 160 m by 53 m;
- spreading of a gravel bed on the sand berm with a thickness of 1 m.

Tolerances applied to the smoothness of the berm structure based on strength characteristics of the caisson were as follows:

- sand: $\pm$ 0.10 m;
- gravel: $\pm$ 0.04 m.

However it transpired that for the prevailing sea conditions it could not be determined with sufficient accuracy whether or not the smoothness requirements were fulfilled. It was decided to place the caisson on temporary support blocks followed by flushing sand underneath the caisson (see 20.4).

20.2.4 Types of equipment and work methods

This island was preceded with the construction of another one with a comparable design concept (Tarsiut), which was constructed with locally available building material. The sand body had to be constructed completely with hoppers. Suitable sand with a D_{50} of 320 μm had to be dredged in a borrow area at a distance of 41 km. Approximately 2.5 million m^3 of sand was placed in the island by three hopper dredgers (Geopotes IX and X and the Hendrik Zanen).

The work method involved the emptying of the hopper by pumping the sand via a delivery pipe to the bottom while the discharge opening was shifted carefully along the perimeter of the island under construction. A small sand dike was constructed on the sea bed with a height of approximately 2.5 m and slopes of approximately 1 : 5. Within this perimeter-dike sand could be bottom discharged. A next dike could be built after that while repeating the same procedure. To obtain the required accuracy, the discharge pipe was shifted using an anchored pontoon.

For the flat finishing of the crest of the submerged berm, involving sand as well as gravel, a cutter suction dredger (Aquarius) was available and cutting could be performed as well as „brushing" (sweeping with fixed cutter).

168

Flushing underneath the caisson was executed by way of a pipeline system within the caisson. The sand (22,000 m^3) was supplied by hopper dredgers and pumped directly into the pipeline system. A bottom protection against erosion was placed around the caisson on the flushed sand. This was executed with the hopper dredger which pumped the gravel over board while sailing slowly alongside the caisson.

20.2.5 *Construction aspects*

The most important difference with the execution of the island Issungnak was the fact that local sand could be used, whereas for Uviluk suitable sand had to be obtained from a great distance. Therefore, for this situation other types of equipment could be considered. In addition it was worthwhile to save on the sand volume and to accept a more costly construction method. A lot of effort was put in controlling the effectiveness of the discharge processes (positioning and discharge results). The problem was that sea conditions for surveying were a controlling factor rather than workability of the dredging equipment.

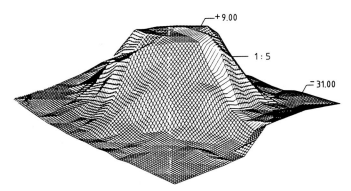

Fig. 78. Three dimensional sketch of Uviluk berm.

The result of the building up of the sand body was that the desired slope of 1 : 5 could be realized but that at various places of the perimeter, slope instabilities occurred after placement. As a result of local flow slides considerably flatter slopes occurred (see Fig. 78). The slope of 1 : 5 was apparently very close to the critical slope under the existing circumstances. The finally used sand quantity could possibly have been less if the construction was initiated at a slope of 1 : 6 and as a result flow slides could have been avoided.

Note:

The next island (Nerlerk), built at a water depth of 44 m, was to a large extent constructed with sand of 270 μm which was locally dredged with a suction dredger. Also here a slope of 1 : 5 was attempted. When the stockpiling had reached a level of −20 m, four large flow slides occurred within one week and further construction with sand was abandoned.

20.2.6 *Feedback with theory*

In Appendix B calculations are made according to theory described in the Chapters 4 to 13. From this the following can be concluded.

The result of the calculations for bottom discharging and single point discharging under water is very sensitive for the spreading width B of the sand-water mixture while flowing across the crater edge and for the spreading width after that (see 5.3). When applying B_{min} at the top of the slope, considerably flatter gradients are derived than those observed. A reasonable agreement is found when applying B_{max}.

According to calculations the flattest slopes while building up an underwater slope, occur during bottom discharging with split barges as was the case for Issungnak. Considerably steeper slopes develop during single point discharging from the water surface and during single point discharging on an abovewater fill area.

Slopes for Uviluk being steeper than those realized for Issungnak, is, apart from use of coarse sand, mainly due to the construction method of single point discharging just above the seabed. However, with this method the risk of flow slide dominating the proces of slope formation is rather large as soon as the sand slope is built up above a certain critical height. According to formula (17) this critical height would be more than 10 m. Due to the fact that no flow slides were observed, it may be concluded that in this case the critical height was much larger.

However, slopes as realized for Uviluk (tan $\alpha = 1 : 5.5$) appear to be at the limit of what can be achieved. They can only be realized with rather coarse sand and a small silt content. This is confirmed from experiences with the island Nerlerk, where large flow slides occurred in a slope of 1 : 5 with sand having a D_{50} of 220 to 260 μm with 5 to 10 % silt or clay after single point discharging just above the seabed [34]. At Uviluk fines could wash away during the suction process; this was not the case at Nerlerk because sand was supplied with a suction dredger.

20.3 Insulating of subsea flowlines, Nessfield and Gullfaks, North Sea, 1988

20.3.1 *Introduction*

On the bottom of the North Sea many pipelines are laid for transport of oil and gas. Some pipelines are laid in a trench, others are laid directly on the bottom.

In 1987 a problem occurred with one of these crude-oil pipelines during pumping of oil.

170

The oil is pumped up at a temperature of 50 °C and cools down in the pipeline to the surrounding water temperature of approximately 6 °C. Contrary to test results, the oil appeared to thicken and a high viscous wax was deposited on the pipe wall.

Out of various possible measures application of an insulating sand layer around the pipeline was selected by the oil company. Because the pipeline was laid in a trench it meant backfilling of the trench up to a guaranteed cover of 0.30 m which had to be protected against erosion with a layer of 0.50 m of rock (see Fig. 79 [54]). Partly due to the success of this project (Nessfield), the same insulation method was applied during a next project (Gullfaks), where the pipeline had to be laid on the sea bed. In order to restrict the sand volume a rubble stone quay was formed on either side of the pipeline creating in a sense a trench on the sea bed (see Fig. 80).

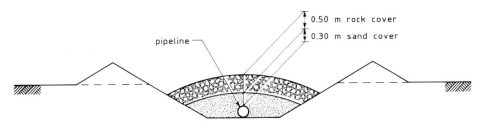

Fig. 79.   Proposed discharge profile at the Nessfield project.

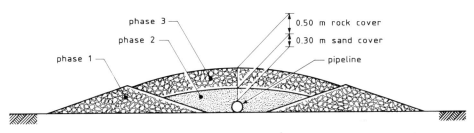

Fig. 80.   Proposed discharge profile at the Gullfaks project.

20.3.2   *External conditions*

Both projects are situated in the northern part of the North Sea. Nessfield is a satellite of the Berylfield at a distance of more than 7 km. Gullfaks is situated in the Norwegian sector of the North Sea. As for Gullfaks, the nature of the sea bed (boulder clay) made trenching unattractive. The water depth at the Ness pipeline is 100 m and at the Gullfaks pipeline 135 m. The workability of the vessels to be used was mainly determined by waves and the fact that operations would be close to offshore structures. The Nessfield project had to be executed during November and December under prevailing bad weather conditions. Anchoring in the vicinity of pipelines and electric cables was out of the question.

20.3.3 *Structure to be made*

The structure to be made consists of bottom discharged quays of sand layers or rock along a trajectory on the sea bed. In Figures 79 and 80 the proposed profiles are indicated.

20.3.4 *Types of equipment and work methods*

Both projects were executed with a special discharge vessel (Seaway Sandpiper), which has the following capabilities:

- dynamic position system;
- loading facility for materials to be discharged;
- vertical discharge bin/fall pipe;
- spider for scanning and monitoring at the lower end of the fall pipe consisting of special cameras, scanning sonars and transponders for the underwater navigation system;
- handling of a remotely operated vehicle (ROV) equipped with cameras, profilers, a pipe tracker, sonars, transponders and responders.

The work method consisted of the following phases:

- in-survey making use of an ROV;
- placement of first quay(s)/layer;
- survey of result using ROV;
- placement of following layer or layers;
- post-survey with ROV.

20.3.5 *Construction aspects*

The execution of the works was governed by the accuracy of process control. The respective elements of process control were:

- input of pipeline position;
- following of this pipeline by the spider at the lower end of the fall pipe;
- tuning of the shift movement along the pipeline according the produced quantity of material.

Without the very extensive continuous registration of progress and results, this kind of discharge operation is impossible. Due to the large dimensions of the vessel (loading capacity 19,000 ton of material), and the capacity of the dynamic position system, operations could continue up to rather bad weather conditions. The total weather down time on both projects remained under 10 %. On these two projects a total of 110,000 tonnes of material (48,000 ton of sand and 62,000 ton of rock) was handled in 65 days. The result of the discharge operation in the Nessfield trench is shown in Figure 81.

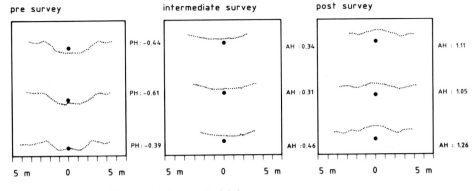

pre survey   intermediate survey   post survey

PH : −0.44   AH : 0.34   AH : 1.11

PH : −0.61   AH : 0.31   AH : 1.05

PH : −0.39   AH : 0.46   AH : 1.26

5 m   0   5 m   5 m   0   5 m   5 m   0   5 m

PH = pipe position with respect to sea bed (m)
AH = cover (m)

Fig. 81.   Result of discharging sand and rock in the Nessfield trench.

20.4   Flushing and pipeline discharging underneath tunnels

20.4.1 *Introduction*

A regularly occurring application of underwater sand discharge is the placement of a sand layer underneath a submerged, temporarily supported, tunnel element. In this case the desired layer thickness is approximately 1 m. The sand layer has a foundation function.

There are two methods for placing the sand layer:

– the „flushing method", at which a sand-water mixture is discharged through fixed, advance mounted injection nozzles in the tunnel bottom [55, 56] (see Fig. 82).

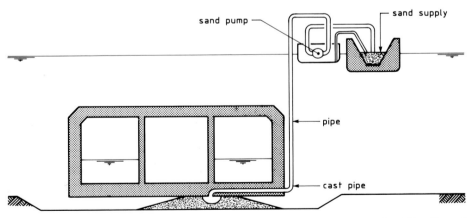

Fig. 82.   Example of the flushing method as applied during the construction of the Hem Tunnel, the Netherlands.

173

- the „pipeline discharge method", at which a sand-water mixture is single point discharged through a horizontal delivery pipe underneath the tunnel; this delivery pipe is mounted on a moveable vertical suspension structure at the tunnel side [56] (see Fig. 83).

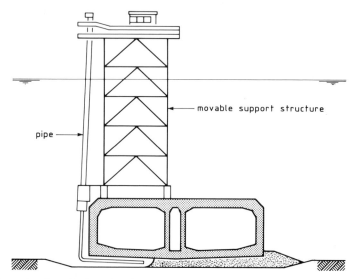

Fig. 83.   Example of the single point discharge method as applied during the construction of the Benelux Tunnel.

20.4.2   *Brief description of applied method*

Flushing method

During flushing, sand is initially pumped with a high concentration to quickly achieve the joining of sand fill against the tunnel bottom. As soon as this is achieved a flat plain is formed against the tunnel bottom in which one or more small „rivers" are formed by the sand-water mixture. These rivers spread and rotate around the injection point as a result of sedimentation and erosion and in this manner distribute the supply evenly along the perimeter of the flat plain shaped sand layer.

The pressure at point of discharge depends on the flow rate, concentration and grain size and especially on the length of the small „rivers" (= radius of the sand body). For reasons of stability (among others, preventing lateral movement) and strength of the tunnel element, this discharge pressure must not exceed a certain value. The total upward pressure force must not exceed 1000 to 2000 kN in order to limit the required ballast in the tunnel.

By adjusting the concentration and choosing the right grain diameter, the required diameter of the „pancake" can be made. When applying a grain diameter which is too coarse, the shape of the sand body becomes a-symmetric as a result of sedimentation

174

effects. Moreover, in that case the chance of sedimentation of sand in the delivery pipelines is also greater. For these reasons in practice the choice of the grain diameter varies between 180 μm and 250 μm. In special situations, such as for the building of breakwaters with caissons, flushing under the caisson can possibly be carried out using gravel.

Generally ample distribution of injection points is applied: centre to centre approximately 10 to 20 m. The end result is obtained as soon as the sand body reaches the edge of the tunnel bottom. This can be assessed by divers.

Pipeline discharge method

During single point discharging underneath the caisson the sand body is formed ahead of the horizontally directed delivery pipe. The final diameter of the sand body depends on the flow rate, the discharge velocity, the grain size and the concentration. With this method coarser sand can be applied because the velocity in the pipeline system and the nozzle can be chosen freely without danger for too large excess pressures under the tunnel.

Apart from verification with divers use can also be made of return suction pipelines which suck directly next to the nozzle. As soon as these suck up a sand-water mixture, the sand body is formed directly ahead of the nozzle and the nozzle can be shifted to a next discharge location.

20.4.3 *Some practical figures*

For both cases the sand-water mixture is prepared in a soil pump installation. The production level varies between 50 m³/hr and 150 m³/hr with a mixture velocity of approximately 4 m/s and a concentration varying between 10 % and 50 % depending on the process phase. The applied grain size diameter varies between approximately 150 μm and 1500 μm, depending on local circumstances. In special cases a stabilizer can be added to increase the bearing capacity of the sand body.

For the consolidation of the pipe-line-placed-sand-layer generally an order of magnitude of 0.05 m has to be accounted for per metre layer thickness of placed material.

20.4.4 *Preparation and trench clean-up*

An important aspect of flushing or pipeline discharging underneath a caisson is cleaning up of the trench bottom. In order to prevent undesired extra consolidation silt layers have to be removed before hand. This especially is a matter of concern in tidal areas where extra siltation may occur as a result of back and forth moving silt density currents. In some cases even mechanic removal of a silt layer is required because this layer can adopt the consistency of soft clay through consolidation [57].

20.4.5 *Hindrance to shipping*

For the flushing method there will almost be no hindrance to shipping.

20.5 Construction of a ring dam in the North Sea, Rotterdam-Europoort, 1986/1987

20.5.1 *Introduction*

The fill disposal area by the name of „Slufter" is situated offshore from Rotterdam-Europoort, The Netherlands, and was constructed during 1986/87 for the storage of contaminated dredged spoil. The disposal area was built as a peninsular in the North Sea at the Maasvlakte coast and consists of an 8 km long, 23 m high ring dam enclosing a 28 m deep pit with a surface of 2.8 million m^2 (see Fig. 84).

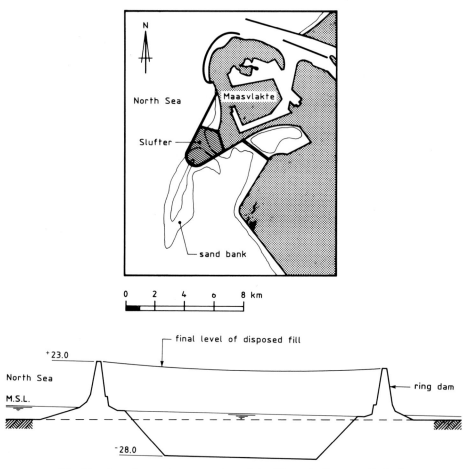

Fig. 84.   Location and cross-section of the fill stockpile area „Slufter".

20.5.2 *External conditions*

The area consists of gullies and sand banks. The depth varies between M.S.L. and M.S.L. +4.5 m (see Fig. 85).
The average local tidal amplitude is 1.91 m (from M.S.L. +1.2 m to M.S.L. −0.71 m).

176

Table 7. Wave data of navigation platform Goeree for the period 1972 to 1982.

wave height ranges H_s (m)	compass sectors (% of occurrence)						
	N 345°–015°	NWI 315°–345°	NWII 285°–315°	W 255°–285°	SWI 225°–255°	total 015°–225°	other sectors + wind still
0.0–0.5	0.95	0.57	1.00	1.09	1.78	5.39	
0.5–1.0	1.80	1.35	1.67	2.39	4.39	11.61	
1.0–1.5	1.16	1.41	1.65	2.13	4.26	10.63	
1.5–2.0	0.61	1.20	1.23	1.88	2.93	7.85	
2.0–2.5	0.30	0.69	0.95	0.99	1.38	4.31	
2.5–3.0	0.20	0.46	0.56	0.59	0.59	2.40	
3.0–3.5	0.12	0.23	0.35	0.35	0.19	1.25	
3.5–4.0	0.04	0.07	0.18	0.19	0.07	0.55	
4.0–4.5	0.04	0.02	0.06	0.07	0.04	0.24	
4.5–5.0	0.02		0.02	0.04		0.08	
5.0–5.5				0.01		0.01	
>5.5				0.01		0.01	
total	5.22	6.00	7.67	9.74	15.63	44.26	55.74

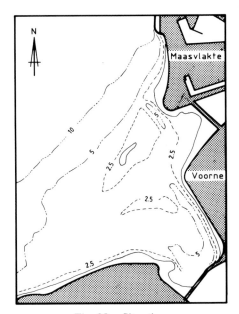

Fig. 85. Situation.

In Figure 86 approximate values are given of the current velocities for average tide conditions. The wave conditions at deep water are given in Table 7, expressed in the significant wave height H_s.

The geological profile of the sea bottom is schematically indicated in Figure 87. Up to 15 to 20 m below sea bed fine sand occurs on top of a layer of hard clay having a thickness of approximately 2 m. Below this clay, medium Pleistocene sand occurs up to approximately 40 m below sea bed.

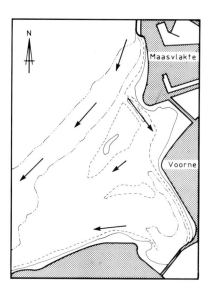

a. 5 hours before high water

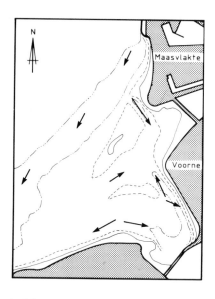

b. 3 hours before high water

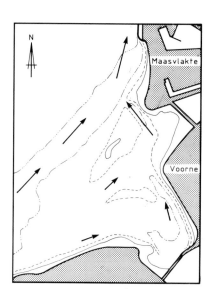

c. 1 hour after high water

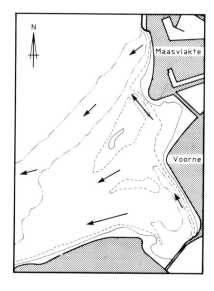

d. 4 hours after high water

0 0,2 0.4 0.6 0.8 1.0 (m/s)

Fig. 86.   Average current velocity and direction for different tidal phases, given for a number of hours before, respectively after high water in Hook of Holland.

178

Fig. 87. Geological profile.

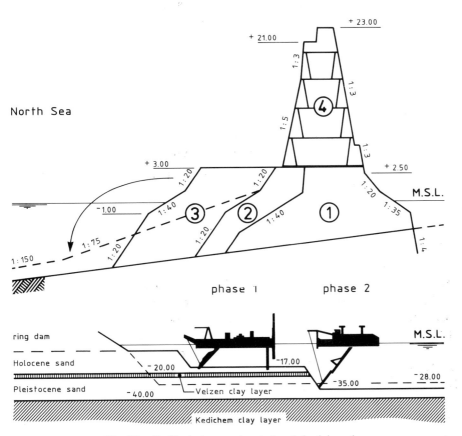

Fig. 88. Profile to be constructed and dredging phases.

179

20.5.3 Structure to be made

In the area a ring dam had to be constructed with a length of 8 km up to a height of M.S.L. +23 m by hydraulic placing of sand. The sand had to be obtained within the ring dam where the suction pipe was allowed to be lowered to a depth of M.S.L. −35 m. Because of environmental considerations the bottom was specified to be at a depth of approximately M.S.L. −28 m.

In the design it was envisaged that the ring dam had to be made with the fine sand as well as with the medium sand from the (borrow)pit. Furthermore at the sea side between M.S.L. +3 m and the original sea bed, a layer of 2.5 m fine to medium sand (D_{50} between 200 μm and 300 μm) had to be placed to increase the resistance against waves and currents of the beach to be constructed. This average diameter would have to be obtained by mixing the fine Holocene sand and the medium Pleistocene sand (see Fig. 88). Moreover, from an optimalization study it appeared that the erosion could be further limited by making a flatter beach slope below M.S.L. −5 m: 1 : 150 instead of 1 : 75 (see Fig. 88). This so-called equilibrium profile should then be in dynamic equilibrium for the existing current and wave conditions. Because such a profile could not simply be made by hydraulic discharging (a slope of 1 : 40 was accounted for above low water and 1 : 20 below low water) the profile was designed in such a way that at the beach ample stock of fine to medium sand was present to obtain the desired flat underwater beach slope by wave and current action (see Fig. 88).

20.5.4 Types of equipment and work methods

The works program of the contractor was based on the construction profile according to Figure 88. With this scheme the following starting points were chosen:

- to keep the quantity of fine to medium sand used as close as possible to the minimum quantity demanded;
- to make the slopes as steep as possible with the fine sand;
- to attain the highest possible production;
- to minimize down time caused by unworkable conditions.

With reference to Figure 88 first a dam had to be constructed up to a level above high water at the inner side of the cross-section of the ring dam using fine holocene sand (1). At the same time the slopes on the sea side, especially in the zone above low water, were to be constructed as steeply as possible in order to minimize the quantity of fine to medium sand. Because it appeared that these steep slopes would not be stable after prolonged action by the sea, the fine sand was covered as soon as possible by medium coarse sand. In order to minimise the time that fine sand was unprotected as much as possible, the fine to medium sand was to be placed in two layers (2 and 3). This did not only limit the loss of fine sand still further, it also reduced the risk that fine sand would get into the zone where 2.5 m fine to medium sand had to be placed. In that case this fine sand either would have to be removed at high cost, or covered over thus increasing the total required quantity of sand.

To optimize the building of the construction profile the contractor assigned a consultant, complementary to his already extensive knowledge and experience in this field, to conduct research on attainable slope gradients. From the results of this study it appeared that damage caused by a single summer storm could be considerable and relatively large in comparison to the damage anticipated if the fine sand remained uncovered throughout the whole summer. Therefore rapid covering appeared to be necessary.

As soon as the fine sand had reached a height of M.S.L. +3.5 m, and was covered with fine to medium sand, the next layer of fine sand was hydraulically placed between bunds (4). These bunds were constructed with bulldozers, making use of already placed fill. They were covered with plastic sheeting to prevent erosion at the inner side and to prevent instability. The thickness of these layers was 4 m. Surplus water was returned to the borrow pit.

During the execution of the project the contractor constantly strived to assure:

– high sand concentrations, especially when dredging fine sand;
– optimal use of bulldozers to prevent formation of channels on the fill area;
– most rapid covering of fine sand with medium coarse sand.

With this method of working double handling of sand could in theory almost be reduced to zero.

To obtain optimal production, the contractor used cutter suction dredgers for the winning of Holocene bedded sand. For this reason the cutter head was placed just below the „Velzen" clay layer so that this layer was broken. The resulting depth after the first dredge excavation was approximately M.S.L. −17 m. With this method of working not only the maximum cut height was reached, but there was constant production when winning the medium sand with the deep suction dredgers (suction pipe 35 m deep). In the fine sand, concentration and production were further increased by using ladder mounted high pressure water jets which promoted active face formation and a high swing velocity of the cutter dredger.

The sequence of the works was described in the contract specifications (see Fig. 89), starting with hydraulic placing of the sub-layers at the sea side. In this way a sheltered location for the dredger was created as soon as possible. The contractor managed to reduce possible down time further by bringing the dredger to the borrow area through channels at the lee side of the shallows (see Fig. 89). In this way a sheltered location for refuge was created behind a sand bank with very little extra dredging as a protection against possible occurrence of summer storms.

For the chosen work method sand losses by waves and currents were limited (see Fig. 86 and Table 7). During construction these losses can be considerable at the fill front (order 10 %). For northerly directed tidal currents this sand was caught in the area north of the pit where later the connection dam was to be hydraulically placed. For southerly directed tidal currents, lost sand was deposited in the alignment of the main dam.

The closure of the ring dam took place at the lee (land) side of the pit during neap tide

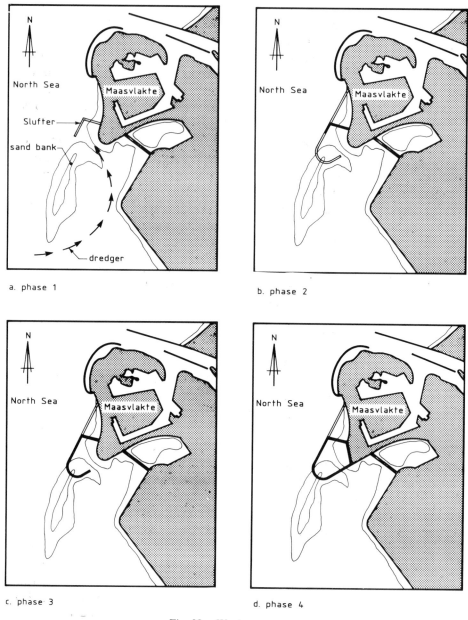

a. phase 1

b. phase 2

c. phase 3

d. phase 4

Fig. 89.   Work sequence.

and no wind conditions. The closure was achieved by single point discharging at the dam head with a capacity of 3200 m^3/hr and with sand with a D_{50} of approximately 300 μm. For this two cutter suction dredgers were employed.

20.5.5 *Construction aspects*

The productions, efficiencies and slopes realized during various phases of the project are indicated in Table 8.

Table 8. Production, efficiency and slopes.

sand/grain size	phase	type of dredger	gross work period (hr)	efficiency (%)	production (10^6 m^3)	slope above water	slope under water
fine sand	I	cutter dredger	5,441	80.0	15.2	1:70	1:30
120–140 µm	I	cutter dredger	4,696	79.2	4.6		
	I	hopper dredger			0.1		
fine to medium	II	suction dreger	6,522	80.9	10.6	1:25	1:15
sand	II	suction dredger	664	84.6	1.7		
250–350 µm	II	suction dreger	1,704	76.0	1.5		
	II	suction dredger	2,695	86.1	3.3		
total/average:			21,722	80.7	37.0		
double handling:					0.5		
dredged net:					36.5		
sand losses:					3.5		
placed net profile:					33.0		

The progress of the work has been assessed on the basis of periodical echo soundings in the excavation pit and on the basis of sub-total quantities mentioned in Table 8. According to specifications the actual quantities of sand handled were determined exclusively from the difference between echo-soundings of the dredged area to be excavated directly before the start of the project and immediately after completion of the total work. The contractor was paid per cubic metre of sand excavated.

According to specifications the client carried out the soundings using their own survey vessel and instruments: the echo sounding frequency for slopes was 210 kHz and for the bottom 33 kHz. This was meant to reduce faulty measurement of the slopes as much as possible in connection with the beam width of the echo signal.

20.6   Construction of a steeper gradient of an existing underwater slope

20.6.1   *Introduction*

The slope of an existing harbour basin, surrounded by reclaimed fine sand fill, had to be provided with a steeper gradient. For this purpose special dredging techniques and additional measures had to be applied.

20.6.2   *External conditions*

The depth of the harbour basin was M.S.L. −13.5 m with slopes of 1:4. Due to increase of draught of the ships the bottom had to be deepened to M.S.L. −19.0 m, resulting in narrowing of the basin between the toe lines of the slope. Placing the slope backwards was not possible because of existing buildings. The local tidal variation reaches 1.6 m and the area is situated at M.S.L. +5.5 m.

20.6.3 *Structure to be made*

A net additional width of 20 to 30 m with respect to the original measure could be obtained by dredging slopes of 1 : 2.5. Before dredging could start a method had to be developed to prevent equilibrium disturbance which could occur during dredging.

20.6.4 *Types of equipment and work methods*

The stability of a sand slope under water is highly influenced by the surrounding ground water. If the piezometric level in the embankments of a channel or harbour basin is higher than the existing water level, a constant ground water flow will occur towards open water. In that case ground water seepage at the slope has to be accounted for. In the considered area the ground water level in the reclaimed fill area varies usually between M.S.L. +2 m and M.S.L. +3 m.

The ground water flow exerts a pressure on the sand in the direction of the flow. This flow pressure is exerted on individual grains and therefore is of great importance for stability of these grains on a slope (micro-stability). In addition to stability of individual sand grains the stability of the ground volume as such had to be evaluated (macro-stability).

From slip failure calculations it appeared that, apart from the mentioned flow pressures, stability of an underwater slope of 1 : 2.5 was still assured. However, the presence of flow pressures required the construction of flatter slopes, namely 1 : 3.5 to 1 : 4. Hence the macro stability appeared to be critical.

The dredging phase is often the most critical phase for the stability of a slope. After all, this will cause a rather sudden change in the stress condition of the underlying sand body. Especially for sands of low densities this may cause instability in the form of a flow slide. Change of stress condition could mean volume reduction of the sand body causing sudden groundwater discharge for saturated sand. For fine sand or for a relatively thin layer of sand between clay layers of low permeability such a discharge is often not possible. The water pressure σ_w will increase and the effective stress σ_g – and hence the shearing resistance – will decrease, which may cause a flow slide often with slopes of 1 : 10 to 1 : 20.

The stability of the individual outer sand grains of the considered slope of 1 : 2.5 would be assured by lowering the piezometric level in the adjacent area in such a way that during dredging an inward directed ground water flow at the slope could be accounted for. In that case also a greater safety of the stability of the entire slope would result which should be attractive during the dredging phase. It was therefore decided to lower the ground water level in the sand formation between approximately M.S.L. and approximately M.S.L. −20 m with deep well pumps. The basis for the calculation of the capacity and distance between the pumps was that at the toe of the slope, at M.S.L. −20 m, the piezometric level of the ground water had to be lower than the lowest water level. After placing of the slope protection the deep well pumps could be removed.

20.6.5   Construction aspects

The considerations mentioned before resulted in placing of 15 deep well pumps with a capacity of 30 m³/hr distributed along the total slope length of 530 m, at a distance of approximately 5 m behind the crest line. In order to break through the horizontal clay layers present in the sand formation, the vertical drainage was applied along the slope in the form of two rows of drains at a distance of 3 m. In this way the underwater slope reach up to M.S.L. −20 m was dredged to 1 : 2.5 without any problems.

The dredging of the slope was executed with a bucket dredger despite the fact that the bulk of the dredging was carried out by a cutter dredger. This was done because it was feared that the suction of the cutter dredger would have disastrous consequences for the steep slope gradient.

During dredging of the slope it appeared that boxcuts sometimes were preserved because of the deep well pumping resulting in local steep slope gradients. This was a drawback with respect to the slope protection to be placed in a later stage. Therefore, corrections were carried out using a grab crane.

20.7   Backfilling of a sinker trench in the Beer Canal, Rotterdam-Europort

20.7.1   Introduction

A sinker pipeline bundle had to be placed in a pre-dredged trench through an existing deep shipping channel. This case history describes how the rising trench sections in the embankments were backfilled with a 1 : 4 slope gradient.

20.7.2   External conditions

The shipping channel has a width of 700 m between the crest lines of the embankments and a depth of approximately 30 m. The slope of the embankment is approximately 1 : 4. The tidal difference amounts to 1.6 m. The current velocities are less than 0.25 m/s. The trench, which was dredged across the channel, had a bottom width of approximately 6 m and slopes of 1 : 4. The rising section of the trench in the embankment measured vertical depths varying between 6 m and 15 m with slopes changing to 1 : 9. This was done to allow the manoeuvering of floating cranes during the laying of the sinker bundle (see Fig. 90).

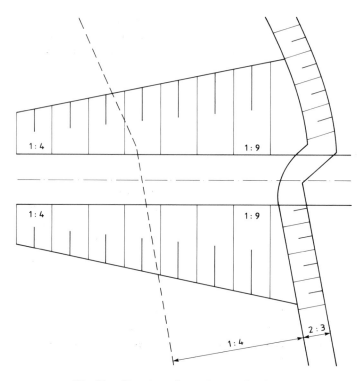

Fig. 90.   Situation of trench at embankment.

20.7.3   *Structure to be made*

The trench had to be backfilled up to the original profile, including the bank sections.

20.7.4   *Types of equipment and work methods*

In order to successfully make a slope of 1 : 4 in sand, excess water pressures had to be prevented in the backfill during placement. If this occurred, failure of the soil mass would result. Because of this the following work method was chosen:

- application of permeable medium sand ($D_{50} = 300$ to 400 μm); this was achieved by washing silt and fines overboard during the loading of the sand in barges in the borrow area, and in this way improving the quality of the remaining sand;
- slow placement of coarse sand in thin layers, and preventing high discharge velocities.

20.7.5   *Construction aspects*

In the present case barges were emptied using a barge unloading dredger and the sand-water mixture was discharged at 2.5 to 3 m above fill level via the suction pipe of a

186

suction dredger. At the end of the discharge pipe a spreader was mounted with a pipe diameter of 0.85 m and a length of 5 m. Here the suction dredger worked in the opposite way and was employed as a spray pontoon.

In a continuous swaying movement parallel to the slope with a velocity of 0.75 m per minute, sand was placed gradually in an upward direction in layers of approximately 1.75 m. As indicated in Figure 91, from the water level to a depth of approximately 14 m, the slope of 1 : 4 was approached rather well. The finishing of the toe to the required slope was done later on by dredging.

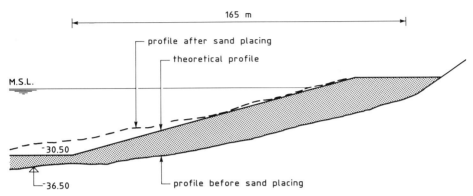

Fig. 91.   Backfilling of trench in the embankment.

20.7.6   *Feedback with theory*

A calculation is given of the three parameters determining the shape of the sand body: the crater circumference, the slope and the sedimentation length (see Fig. 92). Also the crater depth is calculated.

The spreader consisted of a 5 m long horizontal pipe with a diameter of almost 1 m and more than 60 openings at an average distance of approximately 0.3 m. The total area of all these openings was 1.7 m^2; the average opening diameter d_0 was approximately 0.2 m. The mixture flow Q was approximately 1.7 m^3/s. From this it follows that the mixture left the spreader in the form of 60 small jets with an initial velocity U_0 of approximately 1 m/s. The concentration c was approximately 0.15.

The lateral directed jets will have beared off rather quickly under the influence of gravity (resulting downward acceleration taking into account the upward pressure: $9.81(\varrho_m - \varrho_w)/\varrho_m$ = approx. 2.5 m/s^2) and hit the bottom at a maximum horizontal distance of approximately 1 m from the edge of the spreader.

In this way the jets would have hit the bottom over an area of approximately (5 m + 2·1 m) × (1 m + 2·1 m) = 7 m·3 m = 21 m^2 at distances being slightly more than the above mentioned distance of 0.3 m between the openings.

Each small jet would have caused a small crater. With the formula of Breusers (formula (1)) the crater depth of a clear water jet is found: $y_s \approx 1$ m. This is reduced by the

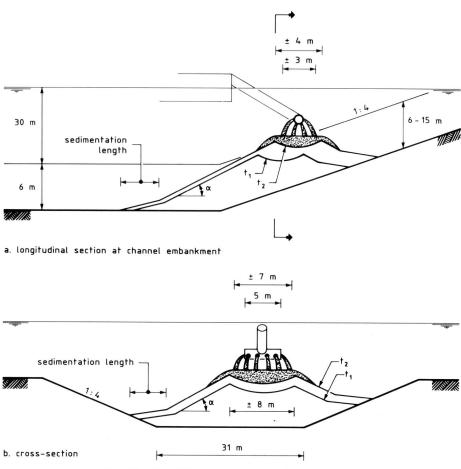

a. longitudinal section at channel embankment

b. cross-section

Fig. 92.   Backfilling of the trench using a spreader.

influence of sand by means of Figure 41: $y \approx 0.2$ m. Also the alternative method of Heezen and Van der Stap (formula (7)) results in this crater depth. The diameter $2r_\infty$ of the small craters is found using formula (6): $2r_\infty \approx 1$ m. From this it follows that the small craters somewhat overlap each other. Probably they formed together one large shallow crater with a length and width being $2r_\infty$ greater than the above mentioned 7 m, respectively 3 m thus giving a crater length of approximately 8 m and a crater perimeter of approximately 24 m.

With the Figures 44 and 45 the minimum width B_{min} is found of the mixture flow passing the crater edge (it is assumed here that $U = 1$ m/s and $d = 0.2\sqrt{60} = 1.5$ m, so that $Q = \frac{1}{4}\pi d^2 U = 1.7$ m³/s): $B_{min} \approx 3$ m. The maximum value B_{max} equals the perimeter of the crater, 24 m. The specific flow rate and the specific sand production rate are calculated with the formulae (13) and (14). The sedimentation length is subsequently found using Figure 48 and the equilibrium slope (without dominant flow slides) by means of formula (21) (see also Fig. 56). The results are given in Table 9.

188

Table 9. Crater dimensions and sedimentation length for spraying just above the sea bed.

magnitude	unit	$B=B_{max}=24$ m	$B=B_{min}=3$ m
specific flow rate q	m^2/s	0.07	0.57
specific sand production s	kg/ms	28	225
sedimentation length L	m	3	20
slope tan α	–	1:3	1:7

The real value of B must have been between B_{min} and B_{max}, so that according to calculations tan α is approximately $1:5$ and the sedimentation length L is approximately 10 m. This corresponds rather well with the profiles found in practice.

20.8 Backfilling of a sinker trench in the Caland Canal, Rotterdam-Europort

20.8.1 Introduction

A sinker bundle has been laid through a shipping channel in a trench at a depth of approximately M.S.L. −31.0 m. In this case history the backfilling of the rising end sections at a gradient of $1:3$ is described.

20.8.2 External conditions

The channel has a depth of 24 m and a width of approximately 500 m. The tidal difference is approximately 1.6 m. The dredged trench has a depth of 7 m in the bottom. The rising section of the trench is approximately 6 m deep.

20.8.3 Structure to be made

The trench had to be backfilled to the original profile with sand having a D_{50} of 300 to 400 μm.

20.8.4 Types of equipment and work methods

For the backfilling a work method by means of vertical spraying was prescribed. One of the requirements was that the opening of the spray pipe had to be kept 2 to 3 m above the sinker bundle.

20.8.5 Construction aspects

Spraying was carried out using a hopper dredge which sprayed the sand through the suction pipe just above the sinker bundle. At the under side of the suction head, through which a sand water mixture was now discharged, a damping plate was suspended to reduce the outflow velocity of the mixture.

Backfilling started with spraying of sand at the bends of the sinker in the rising slopes. After that the horizontal section of the sinker trench was carried out first. Finally the rising sections of the trench in the slopes were sprayed.

However, due to the large outflow velocity of the sand, the sand arrived just outside the sinker trench despite the damping plate resulting in undesired depositions. It appeared impossible to further decrease the flow rate.

Attempts were made to backfill the rising end sections through bottom discharging with hopper dredgers. This failed because the hopper dredger could not be opened gradually. Because of this the hopper content felt as a lump through the water and hit the bottom with force causing craters and mixture flowing partly out of the trench.

In the end the rising end sections were backfilled by means of a pontoon mounted crane. Backfilling started at the top end of the slope. Sand was placed at the water line so that the slope was built up from the top to the toe of the slope (see Fig. 93). Verification took place by means of echo soundings.

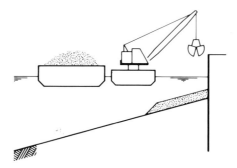

Fig. 93. Backfilling of rising sections of the trench using floating cranes.

20.8.6 *Feedback with theory*

It is interesting to compare this case with the backfilling of the sinker trench in the Beer Canal (see 20.7). Also in this case filling of the trench was initially tried by spraying from 2 to 3 m above the bottom. Why was it successful in the Beer Canal and not here:

- In the first place the slope of the canal embankment should be mentioned: in this case history 1 : 3 and in the Beer Canal 1 : 4, whereas apparently during the backfilling of the sinker trench in the Beer Canal a slope of approximately 1 : 5 was created (the toe was shaped to its final slope by dredging).
- Underneath the vertical pipe a damping plate is applied instead of a spreader pipe with openings. Because of this the outflow velocity of the mixture was much higher and completely in a lateral direction. In this way the bottom may have been hit at a larger horizontal distance from the centre line of the pipe: if the initial vertical velocity was zero and the vertical downward acceleration was 2.5 m/s^2 (a quarter of the gravitational acceleration, corresponding with a mixture density of 1.25 times the

190

water density), the mixture would have reached the bottom after approximately 1½ seconds. In this 1½ seconds the mixture would have covered a distance of 5 to 10 m, assuming an initial horizontal velocity of approximately 5 m/s. As a result the crater would have had a larger diameter: 10 to 25 m. It could be that the crater diameter equalled or was larger than the depth of the trench measured in a horizontal direction, namely $7/(1:3) = 21$ m, so that each building out of the sand body would have arrived outside the required profile.

- The possible mixture flow rate was here possibly also larger than in the previous case (hopper dredger instead of barge unloading dredger), which would have resulted in a larger specific mixture flow rate q and a larger specific sand production rate s. This corresponds with a larger sedimentation length and a flatter equilibrium slope.
- In the description of the working method it is not mentioned that the fine sand and silt fractions were washed overboard during hopper loading, as was the case during the loading of the barges in the Beer Canal. If this would not have been the case the permeability and the effective fall velocity of the sand would have been smaller causing the sand to behave as rather fine sand (see 6.3) resulting in a larger sedimentation length and a flatter slope.

It is not surprising that the trench could not be filled through bottom discharging from a hopper dredger of which the openings could not be controlled. The much larger flow rate and the absence of a damper plate result in much larger crater dimensions. The larger flow rate also results in larger values of q and s and therefore in a larger sedimentation length and a flatter slope. Compare the calculations for Uviluk and Issungnak for bottom discharging with split barges.

20.9 Sand placement for soil improvement under a construction dock dike, River Ems near Leer in Germany

20.9.1 *Introduction*

As part of the Ems tunnel project a construction dock had to be built. This dock was required for the building of concrete tunnel elements. The location of the dock was in the wet lands outside the river dike next to the tunnel alignment (see Fig. 94). The dock consisted of an excavation from +2 m to −9 m with a surface of approximately 200 m × 300 m enclosed by a ring dike to be constructed up to +7.3 m. Because of the expected problems with the stability of this dike soil improvement was carried out underneath and at the inner side of this dike. In the following a description is given of the under water placement of this soil improvement.

20.9.2 *External conditions*

The construction dock dike is situated along the river and forms together with the existing river dike the dock for tunnel element construction. The current velocities in

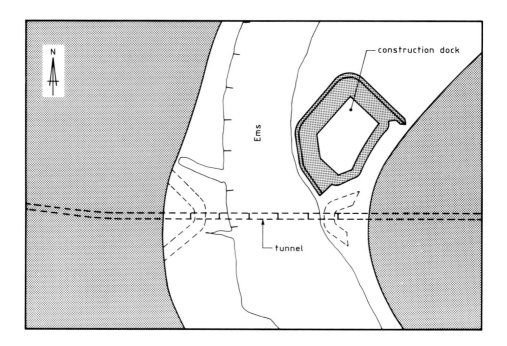

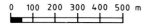

0 100 200 300 400 500 m

Fig. 94. Location of construction dock in relation to the tunnel alignment.

the river vary between +1 m and −1 m/s with a very short turn of the tide. In view of the location close to the coast a normal tidal movement occurs varying between approximately +1.1 m and −1.4 m. Depending on the wind, a water level rise may occur up to 0.5 m at force 6 on the Beaufort scale.

20.9.3 Structure to be made

The design of the soil improvement is shown in Figure 95. During the preparations an underwater slope of 1 : 4 was proposed, based on previous experience. The water depth is approximately 7.5 m.

20.9.4 Types of equipment and work methods

Soft soil with low bearing capacity was removed with a cutter suction dredger by breaking in from the river into the existing wetland. This dredged entrance was closed by hydraulic placement of fill. In the basin formed this way with the surrounding wetland at +2 m, the water level could be controlled with the dredger present. The subsequent soil improvement took place by hydraulic placement of sand with a D_{50} of approximately 250 µm at a discharge production rate of 1250 m^3/hr. The water level was kept at 0.0 m.

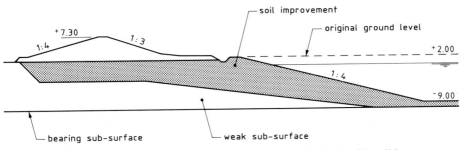

Fig. 95.   Cross-section of the ring dike to be made around the dock with soil improvement.

20.9.5   *Construction aspects*

The underwater slope appeared to adjust to a slope between 1:7 and 1:10. Figure 96 shows a cross-section and plan view of the fill area during pipeline discharging of the first layer at a level of +2.0 m.

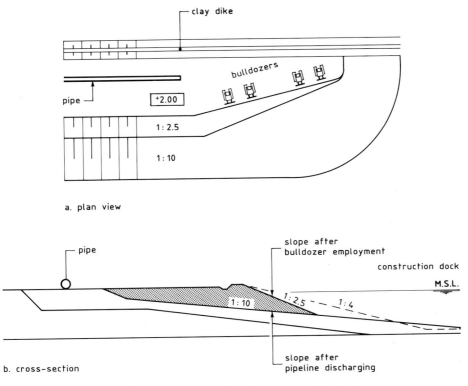

Fig. 96.   Fill area during pipeline discharging of the first layer.

193

In order to achieve the design slope of 1 : 4, sand was moved sideways on the slope with 4 bulldozers (type D6). Over a length of 100 m on the abovewater fill area (from +2.0 m to 0.0 m) sand was either directed to the head of the dam or built out side ways using bulldozers. Fast coupling pipes were used. The obtained slopes were considerably steeper, namely approximately 1 : 2.5.

20.9.6 *Feedback with theory*

It is assumed that the sand-water mixture on the abovewater fill area is pipeline discharged with a concentration c of 0.14 to 0.20. From this a mixture flow rate Q follows of 1.0 to 1.5 m^3/s.

In first instance the mixture was not spread with bulldozers. The mixture might have reached the water level in gullies having a width of approximately 4 m so that the specific flow rate might have had approximately the following value: $q = 0.2$ to $0.4 \, m^2$/s. The concentration would sometimes decrease, sometimes not, so that a variation of c can be expected of 0.1 to 0.2. With this data and by means of Figure 48, the sedimentation length of the underwater slope can now be calculated and the equilibrium slope with formula (21). The results are given in Table 10.

Table 10. Sedimentation under water during single point discharging at abovewater fill area.

magnitude	unit	value
specific flow rate q	m^2/s	0.2–0.4
specific sand production s	kg/ms	60–200
sedimentation length L	m	20–30
equilibrium slope tan α	–	1 : 6–1 : 10

The calculated equilibrium slope corresponds well with the obtained slope. The calculated sedimentation length is considerably smaller than the total slope length (approximately 50 m), but may be larger than the actual occurring length. After all a sedimentation length of approximately 50 % of the total slope length would result in a noticeable flattening of the slope, in any case at the lower reach. Whether this occurred is not quite clear.

20.10 Krammer sill, Eastern Scheldt Estuary, the Netherlands, 1987

20.10.1 *Introduction*

In the Eastern Scheldt Estuary, in the south western part of The Netherlands, a dam had to be constructed. This dam called the Philips dam, was intended to separate the salty Eastern Scheldt water from the sweet water of the planned Zoom Lake and to maintain a sufficiently large tidal difference for the Eastern Scheldt habitat after the completion of the barrier in the entrance of the estuary. Additionally, in this way, a tidal free

194

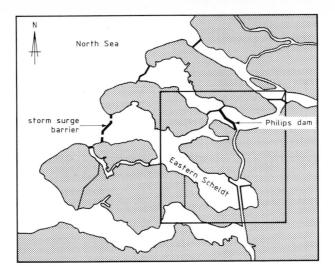

a. Delta area

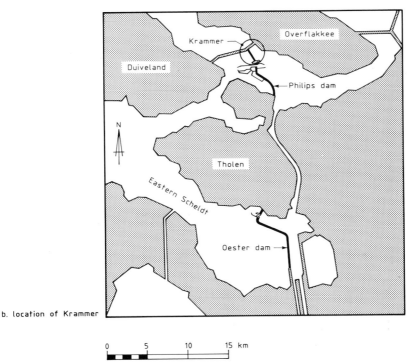

b. location of Krammer

0   5   10   15 km

Fig. 97. Delta area and location of Krammer.

shipping connection could be created from Antwerp to the Rhine river. The location of the Eastern Scheldt in the Delta area in the south west part of The Netherlands is indicated in Figure 97.

Within the scope of the works for the construction of this dam, two tidal channels had to be closed. These closures were carried out using sand. The local depth of the original sea bed at the site was approximately M.S.L. −20 m. As part of the closure of the largest

195

tidal channel, the Krammer, first an underwater berm was made up to a level of approximately M.S.L. −8 m. This berm was the site for the final closure. The closure was mainly achieved by building out from one side in a horizontal direction. The building up of the berm itself is the subject of this case history.

A special feature in the Eastern Scheldt was the possibility of influencing the tidal condition by means of the storm surge barrier located in the entrance to the estuary. Nevertheless this also meant special requirements for the execution of the closure because, for environmental reasons, this tidal reduction was only allowed during a limited period and in a limited manner.

20.10.2 *External conditions*

Configuration

The Krammer channel is situated between the earlier constructed Grevelingen dam and a work island. At an earlier stage a shipping lock complex was built on the work island. As the sand closure progressed, shipping could make use of this.

Figure 98 shows the profile of the closure dam in the centre line of the dam trajectory.

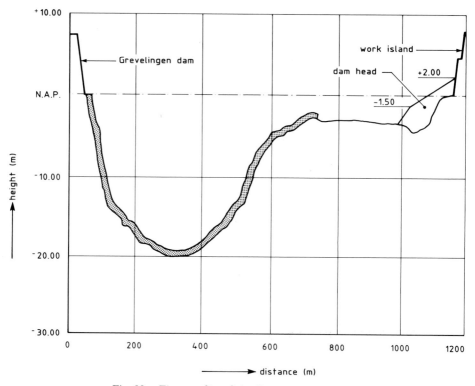

Fig. 98. Flow profile of the Krammer closure gap.

The channel has a total width of approximately 1000 m; half of this is formed by sand banks with a depth up to approximately M.S.L. −4 m, the other half consists of a deep channel with a maximum depth of 20 m.

Sand winning

Sand winning could partly be combined with deepening of the access channels to the lock complex. This sand was used in the lowest part (up to approximately M.S.L. −15 m) of the berm. This sand had a D_{50} of 170 μm. In this way considerable savings were obtained.

Sand for the remaining part of the project had to be dredged at some other location. Nearby to the works two sand borrow areas were found. In assessing the outer boundaries for this the possibility of flow slides had to be accounted for, which could threaten the stability of existing structures or the closure itself. Figure 99 indicates the location of the sand borrow areas. This sand had an average D_{50} of slightly more than 200 μm.

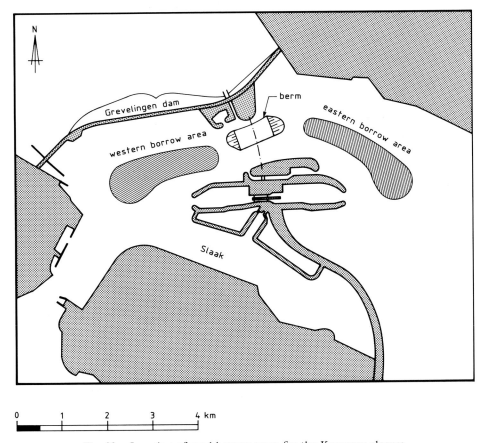

Fig. 99. Location of sand borrow areas for the Krammer closure.

Hydraulic boundary conditions

During the starting phase of the berm construction the storm surge barrier would normally be open. The average current velocity through the closure gap would increase from less than 1 m/s in the original situation to approximately 1.5 m/s at the end of the berm construction with a remaining flow profile of 6160 m². The flow area of the storm surge barrier would then temporarily be limited to 75 % to prevent too much erosion. As a result of this reduction of the flow profile, average high water would be approximately M.S.L. +1.3 m and average low water would be approximately M.S.L. −1.2 m. Due to the sheltered location of the dam section and the borrow areas wave influence was limited. The significant wave height with a frequency of exceedance of 1 % for the duration of the construction, was 0.8 m with an average wave period of 3 s. The large equipment to be employed would suffer little hindrance from this.

Shipping

During the first stage of the closure, the locks in the adjacent work island were not yet in use and therefore unrestricted passage of shipping had to be taken into account during the execution. Also in a later stage, at which shipping could make use of the locks, (anchorage) space for the dredgers in the northerly winning area was still slightly limited.

20.10.3   Structure to be made

During the next construction stage the closure dam had to be built out on the submerged berm. During this stage the storm surge barrier would be closed further which put great demands on the construction programme. For this reason the crest width of the underlying berm had to be larger than the base of the closure dam.
Figure 100 gives a picture of the cross-section of the berm and the closure dam which was to be hydraulically placed on top of the berm at a later stage.
On the basis of this small probability, the expected slopes of the closure dam, the expected erosion of the berm and possible deviations of these expected values, the berm crest was designed. A width of 750 m was selected at a level of M.S.L. −8 m.

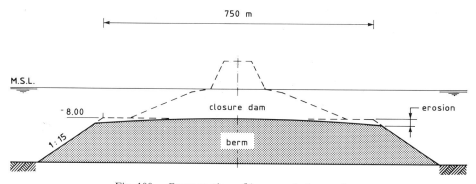

Fig. 100.   Cross-section of berm and closure dam.

It had to be prevented further that instabilities would occur during the building process. For this reason a steepest slope of 1 : 10 was assumed to be acceptable. It was expected that for the large applied production, a slope of 1 : 15 would result.

The total volume of hydraulically placed sand for the construction of the berm would amount to 6.5 million m³, including losses.

20.10.4 *Types of equipment and work methods*

The first stage of the berm construction was achieved using two small dredgers which obtained sand from the access channels of the lock complex.

For the remaining development of the berm two large dredgers were employed, namely the suction dredger Sliedrecht 27 in the western borrow area and the suction dredger Aquarius in the eastern borrow area. As an average during the project the respective productions of these dredgers were approximately 4900 m³/hr and 5200 m³/hr.

The sand-water mixture was delivered from the dredgers to two pontoons in the closure gap via a pipeline system consisting of floating pipelines, land pipelines, a sinker pipeline and again floating pipelines. The floating pipeline in the closure gap was anchored by means of an extra anchor pontoon in order to guarantee the workability during large current velocities.

Figure 101 gives an overview of the production system.

For delivering sand in the berm, a choice could be made between spraying from the water surface or discharging the mixture via a pipe (whether or not provided with

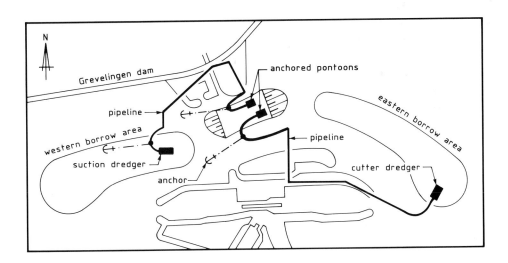

Fig. 101. Production system during berm construction phase.

diffusor) close to the bottom. A choice was made for pipeline discharging without diffusor. This method reduces the losses compared to the work method without pipe, whereas the positive effect of the diffusor was uncertain.

20.10.5 Construction aspects

During the initial phase of the berm construction hindrance from floating ice was experienced. The ice was pushed against the floating pipeline by currents, resulting in forces becoming too large for the anchoring. It was attempted to continue working by connecting the floating pipeline with a break line to the pontoon, in such a way that spraying could be done from the water surface and the pipeline would break away when the forces became too large. At increasing amounts of floating ice, this was not functional and in total three weeks down time occurred as a result.

The berm was divided into four sections and each dredger/pontoon combination was responsible for two sections. First the area west of the centre line of the dam was raised up to approximately M.S.L. −11 m. After that the pontoons had to be switched to a connection on the other side of the channel in order to be able to get to the area east of the centre line of the dam to raise the berm in this area up to M.S.L. −7 m (including some excess height with respect to the desired M.S.L. −8 m because of erosion). Finally the pontoons were switched back again and the western section of the berm was raised up to M.S.L. −8 m. Such a phased execution was considered necessary to prevent differences in height on the berm which could possibly result in irregular and uncontrolled current patterns.

During berm construction care was taken that the pontoons in the closure gap experienced as little hindrance as possible from currents. During periods of large current velocities one worked close to the shore and during small current velocities the area in the middle of the channel was raised.

Because of the accurate work method it appeared possible to develop slopes of approximately 1 : 15. To achieve this for such large productions it is necessary to build up the slope starting from beneath in a step like fashion rather than by just discharging „on top of the mountain".

To obtain a sufficient smooth finish of the berm crest, the work area was divided in sections of 25 m × 25 m. From previous soundings it was calculated how much sand was still required to bring the considered area to the required height. After a pontoon was shifted above the area, the dredger man was notified who subsequently recorded the number of m^3 and gave a signal to the pontoon skipper when a sufficient quantity of sand was placed. After that the pontoon could be shifted to the next section.

Figure 102 shows the final survey of the delivered sand sill.

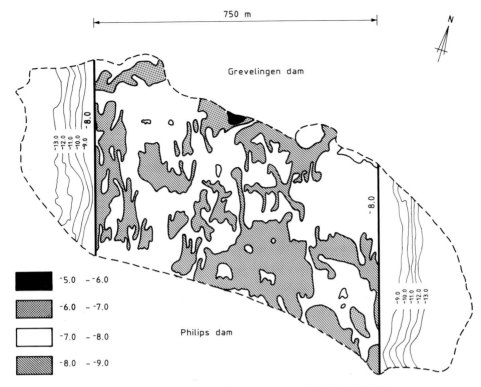

Fig. 102. Final survey of Krammer sand sill (11-4-1987).

20.11 Relocation of a dam, Rotterdam-Europort

20.11.1 Introduction

A dam situated between two harbour basins had to be moved approximately 250 m to the east. Because the water shed had to be maintained the new dam initially had to be placed next to the existing dam before the old dam could be removed (see Fig. 103).

20.11.2 External conditions

On either side of the existing dam there was a tidal difference of 1.3 to 1.5 m. However the phase of the tide on either side lags by approximately 1½ hours.
The depth at the site of the dam to be constructed was approximately M.S.L. −23 m. The eastern flank of the dam was positioned on top of the slope of a local pit with a depth of M.S.L. −35 m. For this reason this pit had to be filled. In the past this pit was used for (temporary) storage of silt and consequently the silt was removed from the bottom and slopes of the pit. Afterwards it appeared that the surroundings of the pit were also covered with approximately 0.3 m of consolidated silt.

201

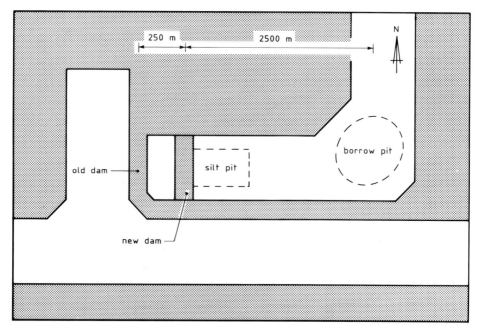

Fig. 103. Situation.

At a distance of approximately 2.5 km a borrow pit was available below the harbour bottom. In order to obtain sufficiently coarse sand, finer sand from above M.S.L. −34 m had to be mixed with medium coarse sand from below this depth (coarser than 300 μm).

20.11.3   Structure to be made

The structure to be made consisted of hydraulic placement of a sand dam. Because of nearby moorings of ocean going vessels slopes had to be made at 1 : 4. The crest width of the dam was 5 m at M.S.L. +5 m and the length was 350 m.

20.11.4   Types of equipment and work methods

In order to be able to construct the steep slopes hydraulic placement in layers was prescribed using a spray pontoon. Otherwise the work method was not further detailed by the client.

For the works the contractor employed a suction dredger with a pumping capacity of 800 hp and a maximum suction depth of 40 m.

The pumping capacity consisted of two boosters with a total capacity of 4000 hp. Pipeline placement of a sub-layer with a thickness of 5 m was placed along the length of the dam starting at one end. This was designed to push ahead and displace the silt present including silt which may result from dredging.

202

20.11.5 *Construction aspects*

During the execution of the works a number of incidents were encountered which are described below.

Shortly after the works had reached a height of approximately 10 m over the full length, a slip failure occurred on the south side. A silt layer must have been present on the original harbour bottom which was not pushed away and which had formed a lens in the dam profile. During further raising of the dam a slide plane developed along this plane. With some soundings and two borings the extent of the lens could reasonably be assessed.

Before continuing the construction it was decided to clear the extruded silt mass and to construct a support berm to prevent further slip failure over any remaining silt lenses in the dam (see Fig. 104).

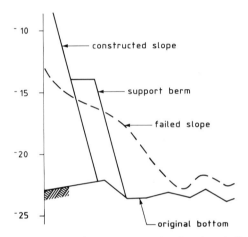

Fig. 104. Slip failure as a result of enclosed silt.

As the work progressed the sand became finer. It appeared that the dredger was unable to pump up the deeper medium coarse sand because of too little pumping capacity, despite the pump being mounted on the ladder. This became apparent after the top part of the sand volume was spent. A solution to this would have been to move 150 m eastward and to open a new borrow pit. However the contractor did not have the required extra floating pipeline.

The result was that the relatively fine mixture did not settle sufficiently and flowed down the already finished slope which was more or less placed at 1 : 4. At the lower reach of the dam slopes of 1 : 7 to 1 : 10 were developed, which had to be dredged to the required profile after completion of the works.

When the dam was raised above high water, bunds were placed in order to raise the crest of the dam to the required height. At this stage the suction dredger was unable to function. Subsequently a suction dredger with a total installed capacity of 3600 hp was hired to complete the works together with a booster of 2500 hp. Even when the dredger

operated at minimum capacity the bunds appeared unable to withstand the considerable force on the narrow fill area. Breaches and again flows down the slope were the
result. At the same time outflowing of sand took place at the down stream end of the
open fill area (see Fig. 105).
This problem could have been solved by hydraulic placing of a sand stockpile of
suitable size and carting sand from here into the crest.

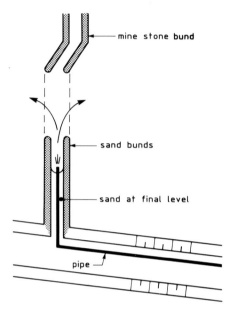

Fig. 105.   Outflowing of sand.

20.11.6   *Analysis*

During this work a higher than normal percentage of drawbacks and calamities were
encountered. Therefore it is worthwhile to analyze what went wrong:

a. In principle the designed structure could be achieved: ample medium coarse sand
   was available (albeit at a large depth) and by following a careful work method the
   prescribed slope could have been achieved.
b. During the preparation of the works it was neglected to carry out a soil investigation
   in the harbour bottom at the site of the new dam and failure to detect the 0.3 m thick
   layer of silt. The presence of the old silt pit should have given cause for investigation.
c. It became evident too late that the employed equipment was unsuitable for this type
   of work. It was impossible to dredge deep enough in order to retrieve sufficiently
   medium coarse sand and not enough discharge pipeline was available to shift the
   dredger to a further located section of the borrow pit.

d. The construction of a dam with slopes of 1 : 4 entails the following requirement:
   – use of medium coarse sand;
   – placement of sand in an even manner with small discharge velocities.
   If sand is directly hydraulically placed into the works with a dredger quality control on sand is only possible afterwards. For the considered case too fine sand was used because of unsuitable equipment. The process could better have been controlled if the dredged sand had been loaded in barges in the winning area. The finer fraction would have disappeared through overflow and after taking samples, suitable sand could have been brought to the works using a barge unloading dredger.

20.11.7   *Recommendations*

With reference to Chapter 19 and Appendix D which deals with quality assurance, it is recommended that the following activities should be developed in order to reduce the risk to an acceptable level:

a. A proper analysis of the operational problems, amongst others investigation of the stability during the construction stages and compliance with specifications especially concerning the final phase.
b. Development of alternatives for the operation (including equipment to be employed) and the choice from these taking into account criteria for design, execution and costs.
c. The application of a risk analyses and the setting up of a manual for the selected construction alternative.
d. Taking care of quality assurance during preparation and execution which includes monitoring of the construction.

20.12   **Dredging and backfilling of a trench for a gas pipeline, Zuidwal project, Waddenzee, the Netherlands**

20.12.1   *Introduction*

The Zuidwal project involved the installation of a 20 inch gas pipeline, a 4 inch glycol pipeline and a 2 inch power/communication cable from Harlingen to the production platform in the Inschot on the Waddenzee north of the city of Harlingen in the northern part of the Netherlands. The length of the alignment was approximately 20 km and the pipelines had to be dug into the shallows with a minimum cover of 1 m and at the location of oyster banks 2 m (see Fig. 106).

20.12.2   *External conditions*

Large sections of the alignment crossed shallows which fall dry at low water. The tidal difference was approximately 2.0 m. The current varied with the location in the alignment and the tidal cycle, the average value however, is approximately 0.5 to 1.0 m/s.

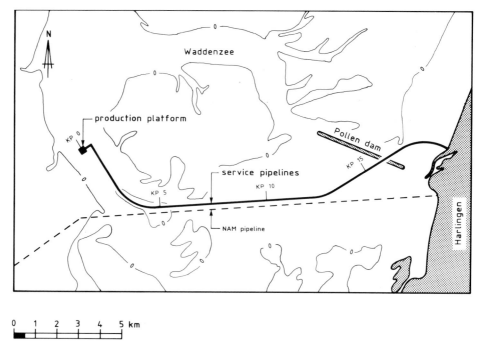

Fig. 106.   Pipeline alignment of Zuidwal project.

In order to disturb the vulnerable Waddenzee habitat as little as possible a minimum profile was dredged i.e. a bottom width of approximately 10 m which was backfilled again.
The sand present had a median grain diameter of 100 to 125 µm with varying silt content.

20.12.3   Structure to be made

The operation was carried out with the so called small scale tandem method. This involves a cutter suction dredger dredging the trench, directly followed by the lay barge which laid the pipes into the trench after which the trench was backfilled with material dredged at the front by the cutter dredger (see Fig. 107).

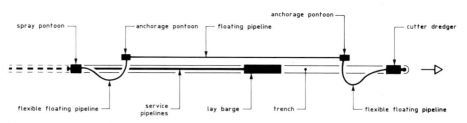

Fig. 107.   Position of equipment for dredging a trench, laying of a pipeline bundle and backfilling of the trench.

206

20.12.4 *Types of equipment and work method*

Sand for the backfilling was obtained by the cutter dredger at the front of the tandem and placed in the trench with the spray pontoon after the pipes were laid.
The sand was placed in a controlled manner through a diffusor suspended from the spray pontoon.

20.12.5 *Construction aspects*

The sand was transported from the cutter dredger to the spray pontoon through a 1300 m floating pipeline. The mixture flow rate was approximately 1.5 to 2 m^3/s with a concentration c of approximately 0.2.
After dredging the trench, the laying of the pipeline and backfilling of the trench, soundings were carried out every 20 m to verify whether the specifications were met. Because of large distances and very shallow sections of the Waddenzee it was decided to equip the cutter dredger „Otter" with extra side pontoons in order to reduce the draft and hence to increase effective working time.
The requirements with respect to the backfilling were that the old sea bed had to be reinstated with a tolerance of ±0.15 m.
In order to meet this requirement a lot of attention was given to the construction of the spray pontoon (see Fig. 108). The pontoon was fitted with a spray pipe with a diffusor

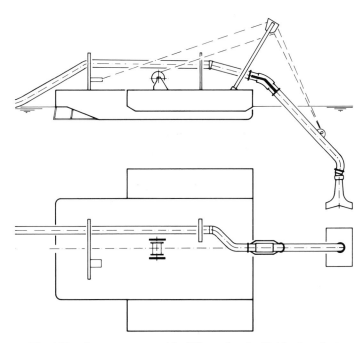

Fig. 108.   Spray pontoon with diffusor for the Zuidwal project.

attached. The diffusor was connected to the spray pipe by means of a ball joint and a bend in order to guarantee a horizontal position of the diffusor. The spray pipe was further equipped with a pendant to determine the depth of the diffusor. The pontoon was also furnished with a positioning system which indicated the position of the diffusor to the skipper on a monitor display.

A survey boat was continuously present with the pontoon to register the progress of the backfilling. The location where the sedimentation in the trench reached the sea bed was 30 to 40 m behind the diffusor and therefore continuous surveying was necessary to obtain a satisfactory result.

Despite low discharge velocities from the diffusor and a proper process control on the spray pontoon and continuous surveying it was not possible to remain within the required tolerances. Therefore at the end of the work some backfilling was required in some locations. This was mainly required by the fact that the soil was not uniform and extensive silt enclosures occurred in the sand which resulted in some losses.

20.13 Construction of the superstructure of a spending beach breakwater, Saldanha Bay, South Africa, 1975/1976

20.13.1 *Introduction*

Saldanha Bay is situated on the Atlantic Ocean, some 110 km north of Cape Town. During the period 1973 to 1977 a harbour is constructed in the bay for export of iron ore and other minerals. The harbour is designed for ships up to 250,000 ton dwt and mainly consists of a spending beach breakwater [58], a loading jetty with access channel and approach dam to the jetty (see Fig. 109).

The construction of the breakwater can be distinguished in two phases: respectively the construction of the substructure and the construction of the superstructure.

In this case history the construction of the superstructure is dealt with during which the construction process was governed by considerable wave heights and wave periods.

Initially it was attempted to create the breakwater without a distinct separation of the construction in two phases. The underwater section would be bottom discharged up to maximum height using hopper dredgers. At the same time the structure would be raised above water by a cutter dredger starting from the peninsular Hoedjies Point and working within its lee. This method was not successful because the initially estimated production capacities appeared to be too optimistic. Subsequently a new work method was designed resulting in a method of construction in two phases as described below. The construction of the superstructure of the breakwater took place from the 30th of November 1975 to the 4th of February 1976. Within 14 weeks the connection of the superstructure with Marcus Island was achieved.

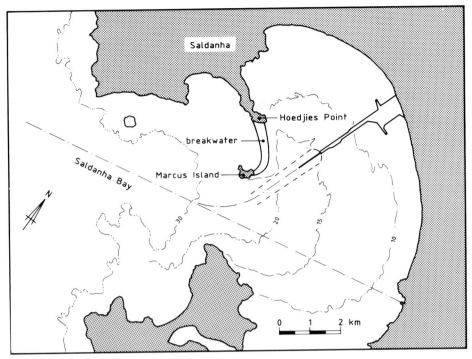

Fig. 109.   Saldanha Bay harbour.

20.13.2   *External conditions*

The average tidal amplitude was approximately 1.0 m and varied between CD + 0.2 m and CD + 1.5 m. The tidal current was small with values up to 0.25 m/s.

The average yearly wind occurrence is indicated in Figure 110. The prevailing wind direction was south west.

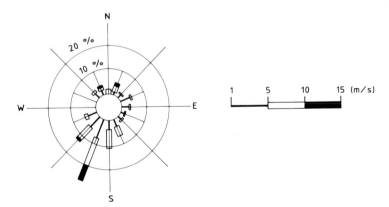

Fig. 110.   Average yearly wind occurrence in Saldanha.

The wave conditions are given as frequency of exceeding significant wave heights H_s as shown on the wave height probability curve in Figure 111, which is based on measurements 1500 m seaward of the breakwater. In this Figure also the average wave conditions at the breakwater are shown.

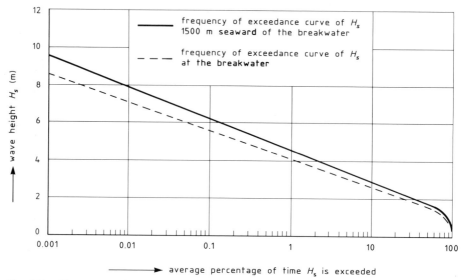

Fig. 111.   Frequency of exceedance of wave height, 1500 m seaward of the breakwater (measured) and at the breakwater.

A summary of the wave periods is given in Table 11. Taking into account the enormous fetch length in the southern part of the Atlantic Ocean, these periods are longer than the period usually encountered on the northern hemisphere. This is combined with a corresponding considerable wave energy impinging on the local coasts.

Table 11.   Average wave period in Saldanha Bay.

frequency of exceedance (%)	T_s* (s)	$T_{s\,av}$** (s)
50	13.4	–
10	–	15.3
5	–	15.7
1	–	16.6

* T_s = period corresponding with the given frequency of exceedance
** $T_{s\,av}$ = average period of exceeded waves

The predominant deep sea wave direction is south west. For all deep sea wave directions the combined effect of diffraction and refraction results in almost identical wave directions at the breakwater. In Figure 112 the wave patterns are indicated at the breakwater for the most frequent deep sea wave directions. In order to assure the lateral

210

stability of the seaward slope of the breakwater, the lay out of the breakwater was chosen as close as possible parallel to the incoming waves using model studies.

In Table 12 the average occurring wave conditions measured during 1970 to 1974 are compared with the wave conditions occurring during the construction.

For interpretation of this Table the large wave periods have to be taken into account. It appeared that the wave conditions during November were less favourable than the average conditions and during December and January these wave conditions were more favourable.

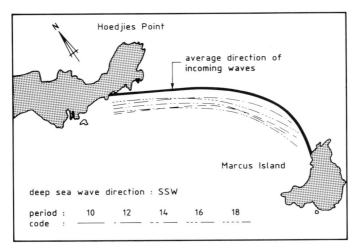

Fig. 112.   Wave patterns according to model studies.

Table 12.   Comparison of the average occurring wave heights with the actually occurred wave heights.

date	average wave height H_s (1970–1974) (m)	occurred wave height H_s summer 1975 (m)
November 1975	1.34	1.43
December 1975	1.35	1.24
January 1976	1.45	1.30

20.13.3   *Structure to be made*

The cross-section of the breakwater is shown in Figure 113. At the seaward side the slope of the breakwater was designed at 1 : 35 with sand of approximately 300 μm. At the bay side a rubble stone protection was applied with stones of 300 to 1000 kg in the outer layer. The length of the breakwater measured along the axis is approximately 1700 m. The local bottom depth varies between CD −16 m and CD −20 m.

The total required quantity for the construction of the breakwater was approximately 20 million m³. Based on the chosen work method the construction of the breakwater

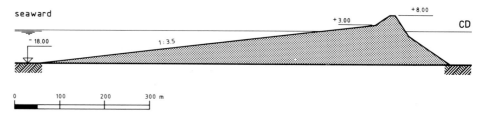

Fig. 113.   Cross-section of spending beach breakwater.

could be divided into the construction of the substructure up to a level of approximately CD −6.6 m for which approximately 15 million m³ sand was required and the construction of the superstructure from CD −6.6 m up to an average level of CD +3.5 m for which approximately 5 million m³ sand was required. The substructure was achieved with the use of hopper dredgers whereas for the construction of the superstructure hopper dredgers were employed equipped with horizontal sliding bottom doors and therefore were suitable to discharge in shallow water. In addition two cutter suction dredgers were employed during this phase.

The final survey of the underwater berm on which the superstructure was built, is shown in Figure 114.

20.13.4   *Types of equipment and work methods*

In order to realize the construction of the breakwater under the given wave conditions and dredgeability of the required sand, a large dredge capacity was present on the site. As part of the construction method of the superstructure two sand stock piles were prepared and in addition to this a discharge pit was used which was prepared in advance. Sand was supplied here by hopper dredgers and subsequently delivered to the works. In Figure 115 the location of the sand stockpiles and discharge pit are shown.

Before starting the superstructure, sand stockpile I was prepared in the lee of Hoedjies Point consisting of 2,000,000 m³ sand. From this stockpile sand was delivered directly into the works by the cutter suction dredger „Queen of Holland".

The cutter dredger „Western Chief" operated from the discharge pit also situated in the lee of Hoedjies Point. This pit was laid out for a capacity of 250,000 m³ per week.

Sand stockpile II was prepared at the lee side of the breakwater as part of the construction of the substructure. At the moment of starting the construction of the superstructure approximately 1,100,000 m³ sand was present. During the construction of the superstructure this sand was replenished with sand which, during construction, was transported outside the limits of the structure to be made as a result of wave action and as a result of bottom discharging with hopper suction dredgers. As the construction progressed this stockpile arrived in the lee of the structure. When sand stockpile I in the lee of Hoedjies Point would have been depleted, the cutter dredger could therefore operate from this stockpile. Change of stockpile took place when the superstructure was built out approximately 800 m.

212

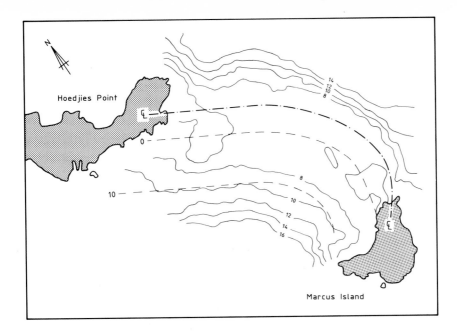

Fig. 114. Final survey of the underwater berm.

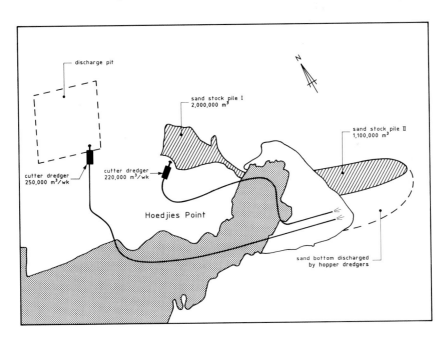

Fig. 115. Situation of discharge pit and sand stockpiles before starting phase 2.

213

The basis for the design of the stockpiles, discharge pit and selection of cutter dredgers was that a continuous production capacity of approximately 500,000 m³ per week could be guaranteed.

Apart from delivering sand with cutter suction dredgers, sand was bottom discharged at maximum level on the underwater berm and on the seaward slope of the breakwater using hopper dredgers with restricted draft and equipped with sliding doors in the bottom (see Fig. 116).

Fig. 116.   Construction of superstructure.

20.13.5   *Construction aspects*

The construction of the superstructure was started from Hoedjies Point. The construction progress is illustrated in Figure 117.

Part of the construction method was the construction of a relatively small rubble stone quay, which was placed at the low water line of the future beach profile as the structure expanded from Hoedjies Point. The crest height of this dam was approximately CD +4.5 m. This dam appeared highly effective: sand was transported out of the seaward profile to the head of the dam by wave action and subsequently settled in the lee of this quay. Here the sand was pushed by bulldozers against the back side of the rubble stone quay. Sand also settled just in front of the quay head and as soon these deposits had reached the low water level, rubble stone was placed and in this way the quay moved forward.

In total more than 235,000 ton of stone of 0 to 4000 kg (quarry run) was used in the quay. For a total length of 1900 m of this quay (including approximately 200 m in front of Hoedjies Point) which is approximately 120 t/m. Due to wave action occasional stone usage was 300 t/m.

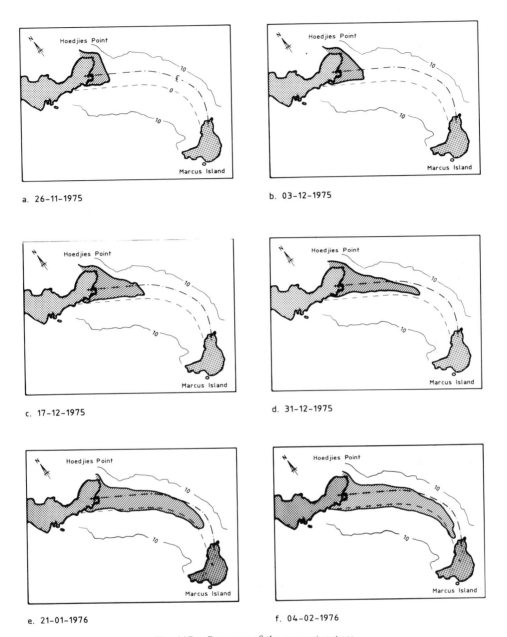

a. 26-11-1975

b. 03-12-1975

c. 17-12-1975

d. 31-12-1975

e. 21-01-1976

f. 04-02-1976

Fig. 117. Progress of the superstructure.

During construction, on two occasions, waves occurred with an H_s of 3.5 m and 3.3 m, respectively with maximum values of approximately 5 m. During one of these occasions the head of the quay was set back some 40 m.

The pipeline plan recognised the possibility that sand could be discharged at the sea side of the rubble stone quay as well as at the lee side of this quay. The distribution of the cutter production was respectively approximately 80 % at the sea side and approximately 20 % at the lee side of the quay. This distribution illustrates the dominant influence of waves on the formation of a sand body. As a matter of fact sand placement at the lee side of the quay was determined by the necessary supplement to the sand body to obtain a sufficiently large and safe work area.

Up to 400 m from the head of the rubble stone quay, the seaward discharge pipe had outlets at distances of approximately 100 m. The distance to the head of the quay of the most forward outlet of this seaward pipe, varied between 50 m and 100 m. During high waves the more backward situated outlets were used and during less wave action sand was discharged at the head of the quay. In most cases sand was discharged at a distance of 150 m from the head of the quay.

The slope achieved with this work method approached the design slope of $1:35$.

Behind the quay, sand was also placed in the structure. The distance between the water line and the pipeline outlet varied between 100 m and 200 m with an average of 125 m. The pipeline level average was CD +3.5 m. The width of the fill area was 40 to 50 m. The pipeline diameter of the „Western Chief" was 0.75 m and the average flow rate was approximately 1.8 m^3/s with a concentration of approximately 0.2. The average grain size was 370 μm.

The pipeline diameter of the „Queen of Holland" was 0.9 m and the average flow rate was 2.5 m^3 with a concentration of 0.15. The average grain diameter of the sand in stockpile I was 640 μm.

As refered to previously sand was also delivered directly into the works with shallow drafted hopper dredgers, mainly directly in front of the progressing superstructure and in the seaward profile to obtain a proper connection with the profile already realized with the cutter dredgers. In this way the level of the substructure was raised from CD −6.6 m to an average of CD −5.15 m.

With the different types of dredgers the following quantities were placed into the super-structure:

hopper dredgers:	2,000,000 m^3
cutter dredgers:	5,100,000 m^3
total:	7,100,000 m^3

The quantity required between the final
survey of the substructure and the superstructure
at a level of CD +3.5 m was: 5,600,000 m^3

Thus the sand losses during the construction
of the superstructure were: 1,500,000 m^3

This is 21.1 % of the total delivered quantity of sand during phase 2.

Loss of sand is defined here as sand deposited outside the contours of the structure. However most of this sand deposited in the planned sand stockpile II situated at the lee side of the structure and was taken up again and placed in the works by the cutter dredger during the progress of the superstructure.

Except sand replenishment from „losses" from the superstructure, sand was also placed in sand stockpile II by hopper dredgers during the construction of the superstructure. This quantity was approximately 700,000 m^3.

CHAPTER 21

CASE STUDY

CONSTRUCTION OF AN UNDERWATER DAM FOR A CONSTRUCTION DOCK
FOR A CONTAINER TERMINAL IN ROTTERDAM-EUROPORT, 1988

21.1 Problem definition and situation

21.1.1 *Problem definition*

In a harbour basin in Rotterdam-Europort a quay wall had to be constructed in a construction dock. The existing sea bed varied between M.S.L. −14 m to M.S.L. −24 m. At the water facing side the boundary of the construction dock consisted of a sand dam which was up to M.S.L. −7 m reclaimed by hydraulic placement of sand under water. The second phase, the raising of the dam up to the required height from M.S.L. −7 m up to M.S.L. +4.5 m, was carried out in the dry. For this, sand was used from the excavation of the dock. The work was executed from March 1988 until September 1988.
In this study the construction of the underwater dam is described.
The available space for the construction of the sand dam was limited because the outer limits for shipping adjoined the sand dam. In order to prevent the toe line of the sand body overlapping with the turning basin, the slope of the construction dock was designed as steep as possible. It was proposed to win the required sand for the sand dam nearby from a created borrow pit in the harbour.

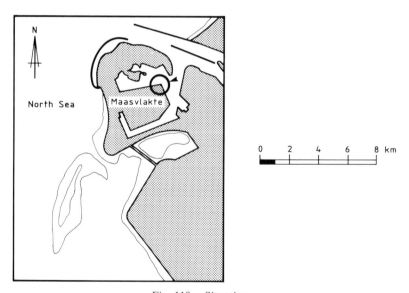

Fig. 118.   Situation.

218

21.1.2 *Situation*

Figure 118 shows the situation of the area. The location where the underwater dam is made is indicated in Figure 119 in relation to the direct surroundings. Figure 120 shows a lay-out and cross-section of the dam to be constructed.

21.1.3 *External conditions*

Tide
The harbour basin has an open connection towards the sea. The tidal difference at the site of the dam to be constructed amounts to 1.70 m (M.S.L. +1.05 m to M.S.L. −0.65 m).

Current velocities
At the site of the underwater dam to be constructed, the maximum tidal current varies between 0.1 m/s and 0.2 m/s.

Wave climate
The wave climate (see Table 13) is determined by wave penetration from open sea as well as by locally generated waves (fetch length approximately 2 km). The data in this Table are derived from wave observations performed during a representative period in the immediate surroundings of the dam to be constructed.

Table 13. Wave climate.

wave height H_s (m)	percentage of time that the wave height H_s is exceeded (%)
0.4	6.7
0.5	3.3
0.6	1.6
0.7	0.7
0.8	0.2

Wind climate
The wind climate is given in Table 14. The data in this Table are derived from measurements of wind speeds performed during a representative period in the immediate surroundings of the dam to be constructed.

Table 14. Wind climate.

wind force (Bft)	percentage of time that the wind force is equal or greater than the indicated force (%)
5	27.2
6	11.1
7	3.5
8	1.0

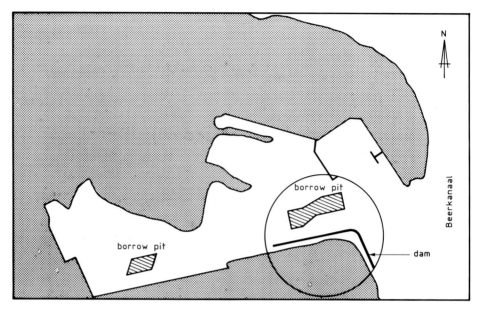

Fig. 119. Lay-out of the dam to be constructed.

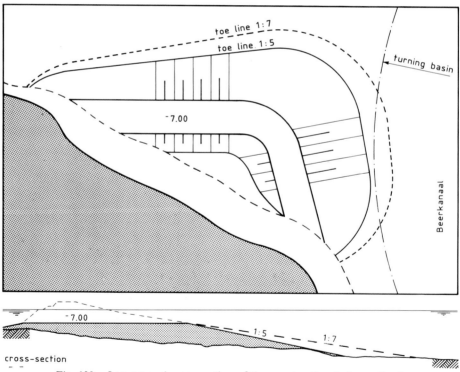

Fig. 120. Lay-out and cross-section of the construction to be realized.

21.2 Geotechnical pre-survey

21.2.1 *Purpose of pre-survey*

Before starting the construction of the sand body a geotechnical pre-survey has been carried out. The purpose of this pre-survey was to indicate in which way a stable slope could be made. The survey was twofold:

a. To determine the suitability of the sand in the borrow area. This suitability was related to:
   - construction of the slope gradient;
   - stability;
   - sensitivity to flow slides.
b. Investigation of the soil condition under the future dam:
   - determining the expected consolidation;
   - determining the stability of the sub-soil and possible necessity for soil improvement.

21.2.2 *Available grain size distributions*

The records of borings already performed in the past in the vicinity of the possible borrow area, were collected in order to obtain a picture of the available grain size distributions of the Pleistocene sand. It was expected that these data would provide a reasonable picture of the sand in the borrow area.

Of 18 sand samples, average sieve values were calculated with the following result (see Fig. 121):

- $D_{60}/D_{10} = 2.1$
- $D_{50} \quad = 350$ µm

These sand samples were obtained at depths varying between M.S.L. −45 m to M.S.L. −51 m. These depths correspond with the suction depth in the borrow pit.

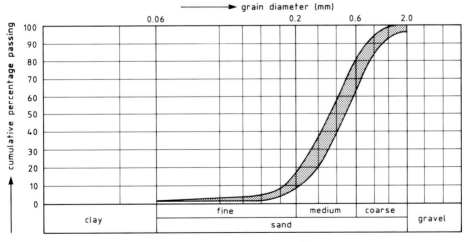

Fig. 121.   Grain size distributions of the sand.

21.2.3 Soil investigation

The soil investigation at the future location of the underwater dam consisted of 9 Dutch cone tests and 3 borings with undisturbed sampling. In Figure 122 a representative sounding is given. On selected samples classification tests, consolidation tests and triaxial tests were carried out.

In 21.2.4 a brief summary is given of the most important recommendations as a result of this soil investigation concerning the method of execution and selection of the sand to be used.

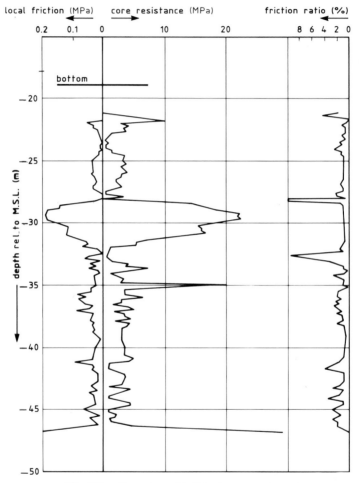

Fig. 122.   Representative Dutch cone test result.

21.2.4 Geotechnical advice based on soil investigation

It was expected that the porosity of the sand in the dam would respectively be 45 % or 48 %, depending whether sand would be bottom discharged or single point discharged.

222

The critical porosity n_{crit} was estimated at 42 % to 46 %. This would mean that the sand would have such a loose condition that it would be sensitive to liquefaction over a height of more than 10 m. According to 9.3 a gradient of the outer slope steeper than 1 : 3 to 1 : 7 would probably result in a flow slide.

In view of the possible chance of flow slides a slope gradient of 1 : 5 to 1 : 7 was advised at the time.

It was also recommended that the present silt layer, having a thickness varying between 0.5 m and 2.0 m, was removed before placing the sand dam.

If hydraulic placement was not to be applied in thin layers, it was expected that within the placed sand layer, an excess hydrostatic pressure would develop of 15 to 20 % of the placed load. This applied to the relative fine material with a D_{50} of 150 to 250 μm. The sand with which the dam was to be constructed should therefore preferably have a minimum D_{10} of 150 μm and a minimum D_{60}/D_{10} of 2.5 (this would imply for the D_{50} a minimum value of 280 μm, see contract specifications in 21.4.1).

These requirements for the sand to be used were on the one hand dictated by the drainage capacity of the sand (or by the permeability which is mainly determined by the smallest grains, hence by the D_{10}) and on the other hand by the compaction characteristics (greater compactability for a flatter grading curve). The drainage capacity had to be such that little or no excess hydrostatic pressure could develop because this would adversely influence the slope stability. The available grain size distribution curves comply with the first requirement but not with the second. However, the sand was considered to be suitable (see also the conclusions at the end of 21.3, the contract specifications in 21.4.1, the test results in 21.5.4, the laboratory investigation conducted afterwards in 21.6.2 and the evaluation in 21.7).

From calculations it followed that the expected settlements as a result of consolidation, would vary between 0.40 m and 1.00 m. In view of the area across which these settlements would occur, no stability problems were expected as a result of differential settlements. These settlements would be reached within 1.5 to 2.5 years after placing the surcharge.

However, the first layers would have to be placed with a small fill height and a small layer thickness in order to obtain the lowest possible excess hydrostatic pressure. Above water level, raising was possible with finer material. For the construction of the construction dock the inner slope could be excavated at a gradient of 1 : 2 to 1 : 3.

21.3   Method of execution

21.3.1   Alternatives and final choice

In view of the experiences with the Delta Works (see 8.5) and in the Beaufort Sea (see 4.5, 20.1 and 20.2) and the results of calculations (see Appendix B) it was anticipated that the required steep slopes could be achieved with pipeline discharging under water, where the pipe mouth would be kept just above the bottom, where a rather small mixture flow rate would be chosen and where the mixture would be distributed and slowed down by means of a T-section mounted at the end of the discharge pipe. In this way the developed craters would also be small. If required, the crater depth would further be reduced by shifting the pipe continuously.

21.3.2   Calculations of the crater dimensions and the slope gradient

The three parameters determining for the shape of the sand body: the crater width, the slope gradient and the sedimentation length are given in Figure 123 and can be calculated with the method described in this manual. The crater depth can also be calculated for the situation that the pipe is kept at the same place during a longer period. During rapid shifting the crater becomes so shallow that it will more appear like a horizontal berm. The calculations are performed for two grain diameters ($D_{50} = 200$ μm and $D_{50} = 400$ μm) and for two configurations of the pipe mouth (one without T-section and one with a T-section).

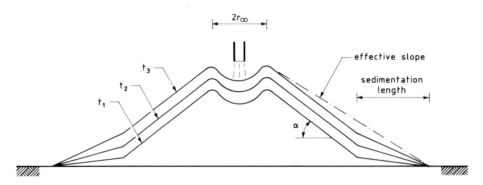

Fig. 123.   Parameters determining for the shape of the sand body.

It is assumed that the works are carried out with a discharge pipe with a diameter d_0 of 0.6 m, a mixture velocity U_0 of 4 m/s and a concentration c_0 after leaving the pipe mouth of 0.2. The height H of the pipeline outlet above sea bed is 3 m. When applying a T-section it is assumed that this section is 5 m long and that the discharge velocity is reduced to 25 % ($U_0 = 1$ m/s). The outlet is schematized as a 5 m long narrow opening. For the same mixture flow rate the width of the opening b_0 is then equal to 0.23 m.

Without T-section
For the situation without T-section the jet which causes the crater will have a round shape. Therefore the formula of Breusers (formula (4)) can be applied to find the crater depth y_s for a clear water jet. This is reduced for the influence of sand by means of Figure 41. Also the alternative method of Heezen and Van der Stap (formula (7)) is applied. No jet calculations were carried out. Because the fall height is relatively small ($H/d_0 = 5$) and the sand-water mixture is pipeline discharged, it may be assumed that the velocity, the concentration and the jet diameter will hardly change over this height.

224

The crater diameter $2r_\infty$ is found by means of formula (6). As described earlier the crater depth will become shallower than calculated here if the pipe is shifted during discharging. The crater diameter, however, remains more or less constant.

By means of the Figures 44 and 45 the minimum width B_{min} is found of the mixture flowing over the crater edge. According to formula (10a) the maximum value B_{max} is equal to the perimeter of the crater. With the minimum value of the overflow width and with the formulae (13) and (14), the specific flow rate q and the specific sand production rate s can be subsequently calculated.

With T-section

For the situation with T-section, in first instance, the jet is considered as a plain shaped jet so that the formula of Rajaratnam (formula (5)) can be applied to find the crater diameter resulting from a clear water jet. The other parameters are found in the same way as for the round jet.

For finding B_{min} with the Figures 44 and 45, for „d" the calculated diameter $\sqrt{0.23 \cdot 5 \cdot 4/\pi} = 1.2$ m is used. According to formula (10b) $B_{max} = 2 \cdot 5 + 2\pi r_\infty$ m.

The sedimentation length is subsequently found by means of Figure 48 and the slope gradient (without dominant flow slides) by means of formula (21) (see also the Figures 56a and 56b). Finally a check on whether the dimensionless parameters (see 8.4) are smaller or equal than 1 is carried out to verify whether or not flow slides will dominate the slope formation. Flow slides occur if $L^* < 1$, $N^* > 1$ and $H^* > 1$. Of these parameters always $N^*(= n/n_{crit}) > 1$, as mentioned in 21.2.4. The parameters L^* and H^* follow from the slope height of 7 to 17 m and from Figure 55.

All the results are summarized in Table 15. It appears that within each column of this Table a large spread of the results occurs. This is caused by the large difference between B_{min} and B_{max}. If no T-section is applied, even spreading of the mixture across the whole crater perimeter of 43 m or 46 m seems unlikely. In that case it is assumed that channel formation occurs with large sedimentation lengths and flat slopes as a result. When applying a T-section, probably slightly more spreading will occur and smaller sedimentation lengths and steeper slope gradients may be expected.

Based on this it is concluded that when applying sand with a D_{50} of 200 μm initially slopes will develop of approximately 1:12. However in this case flow slides will probably dominate, resulting in even flatter slopes. When medium sand is used gradients of 1:6 to 1:7 will be obtainable. From calculations it further appears appropriate to apply a T-section with a length of at least 5 m with which a sligthly steeper slope may be achieved.

In view of risk of flow slides during hydraulic placement or thereafter (see 2.3), it is recommended that the slope gradients are not constructed steeper than 1:5.

For raising the dam from M.S.L. −7 m to M.S.L. +4.5 m another work method was selected. This part was carried out in the dry. For this sand was used from the excavation of the construction dock.

Making a steeper slope will result in considerable cost savings because less sand is required. Therefore the dredging away of the dam around the temporary dock will be less costly.

Table 15. Calculation results of the crater dimensions and the slope gradient for two grain diameters by a pipe mouth without and with a T-section.

magnitude	unit	without T-section $U_0=4$ m/s, $d_0=0.6$ m, $H/d_0=5$		with T-section of 5 m long $U_0=1$ m/s, $b_0=0.23$ m, $H/2b_0=6.6$	
		$D_{50}=200$ μm	$D_{50}=400$ μm	$D_{50}=200$ μm	$D_{50}=400$ μm
y_s (Breusers)	m	9.7	9.0	4.7	3.3
y (y_s reduced)	m	2.3	2.1	1.1	0.8
y (H. en V.d. St.)	m	1.1	1.1		
$2r_\infty$ (Breusers)	m	11.5	10.4	5.7	4.0
$2r_\infty$ (H. en V.d. St.)	m	5.5	5.5		
B_{max}	m	46	43	28	23
B_{min} (Fig. 44)	m	2.9	2.5	2.9	2.5
B_{min} (Fig. 45)	m	2.1	2.1	2.1	2.1
q	m²/s	0.02 to 0.54	0.03 to 0.54	0.04 to 0.54	0.05 to 0.54
L	m	4 to 70	2 to 30	7 to 70	3 to 30
s	kg/ms	13 to 285	14 to 285	21 to 285	26 to 285
slope gradient according to formula (21)		1:4 to 1:14	1:2.5 to 1:7	1:5 to 1:14	1:3 to 1:7
$L*$		0.05 to 0.71	0.05 to 0.61	0.07 to 0.71	0.06 to 0.61
$H*$		1.4 to 3.4	0.5 to 1.1	1.4 to 3.4	0.5 to 1.1

21.4 Contract specifications and tender procedure

21.4.1 Description of contract specifications

The most important contract specifications concerning the working method are as follows:

- The sand has to be reclaimed by means of a suction dredger equipped with a submerged pump. The minimum suction depth is M.S.L. −45 m.
  Explanation:
  Because fine sand is present above M.S.L. −20 m and medium sand below M.S.L. − 20 m, dredging had to be done at least up to M.S.L. −45 m in order to obtain a sufficiently coarse mixture. For this a submerged pump is required.
- The sand has to be transported in barges.
  Explanation:
  The distance between borrow area and fill area was such that it was possible to work with a floating pipeline. However this method was not selected on purpose because the sand quality could not be judged in advance and therefore selection of sand could not take place. When loading barges, the quality is improved through overflow of the fine fraction.

226

- The hydraulic placement of sand has to take place by means of a barge unloading dredger coupled to a spray pontoon. The spreader has to be equipped with a positioning system. The sand-water mixture has to be brought up to approximately 3.0 m above the bottom by means of a vertical pipe. The water facing slope gradient must be between 1:5 and 1:7 and the top side of the sand body has to be at M.S.L. −6.75 m (+ or −0.25 m).
- The fill must be placed evenly and with a constant shifting velocity across the whole area in layers with a maximum thickness of 2.0 m.
- The reclaimed sand will be judged on the quality of the grain size distribution. When measuring in the barges, surface samples will be taken in the barges at different locations which will be put together to one mixed sample. The mixed sample will be determined by means of a fall tube (see 21.5.4) where the D_{50} must have a minimum value of 280 μm.

21.4.2   Tender procedure

It was a negotiated contract because the selected contractor had the opportunity when the sand was rejected by the client, to use this for another purpose. In this way the client incorporated only that sand which accorded with the contract specifications.

21.5   Verification measurements during hydraulic filling

21.5.1   General

During the construction (see 21.5.2) manual soundings as well as echo soundings were performed regularly. Furthermore a sample was obtained from each supplied barge. By means of these verifications (see 21.5.3 and 21.5.4) it was determined whether the required slope was achieved and whether the supplied sand was in accordance with the prescribed grain size distribution.

21.5.2   Construction method

The dam to be constructed is hydraulically placed in layers of approximately 1 to 3 m in order to accelerate water discharge and to minimize the chance on excess hydrostatic pressure and hence the sensitivity to liquefaction (see Fig. 124). The sand was sucked from the barges and subsequently placed under water via a spreader suspended approximately 3 m above the bottom. This height was maintained because otherwise crater formation would occur.

The spreader consisted of a T-section mounted at the end of the delivery pipe. This T-section was closed on one side and provided with a number of openings underneath so that the sand-water mixture could be properly distributed across the slope.

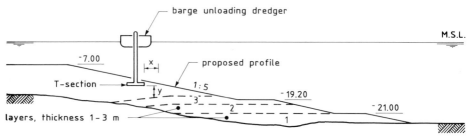

Fig. 124. Construction method.

The horizontal distance x of the T-section to the final slope depends on the grain size distribution of the sand.

For fine sand x increases (up to ≈ 5 m) because this will result in flatter slopes (see also Fig. 124). For coarser sand the closed side of the T-section can be kept more or less at the level of the final slope ($x =$ approximately 0).

21.5.3 *Surveys*

Two types of surveys were performed:

a. By means of a hand lead during hydraulic placement.

   With this it could be determined exactly how the slope gradient developed. When deviations occurred it could then be decided whether or not to shift.

   Shifting parallel to the slope was done the moment the layer had reached a thickness of approximately 1.75 m. The layer thickness was measured by means of a hand lead approximately once every 5 minutes. In this manner the process was well under control.

b. By means of a survey vessel.

   After shifting, the constructed slope was measured by means of a survey vessel so that the extent that the required profile had been achieved could be assessed. After the construction was completed another two surveys were performed with an intermediate period of 3 months (on the 25th of August 1988 and on the 24th of November 1988). Between these two surveys no major differences were observed (+ or −0.10 m).

21.5.4 *Testing of sand in the supplied barges*

A sand sample was obtained from each barge loaded and D_{10}, D_{50}, D_{60} and D_{80} were determined by means of the fall pipe test. Because large grains have a higher fall velocity than the smaller ones, the equivalent grain size distribution can be found rather rapidly by means of the fall test (see Fig. 125) on a sand water mixture provided the fall velocity of a grain of a certain size is known beforehand.

Barges loaded with too fine sand, were rejected so that the quality of the used sand was always in accordance with the required specifications.

The condition that the D_{50} of samples taken from the sand surface in barge always had to be greater than 280 µm, was satisfied for 90.7 % of all controlled barges.

The D_{10}, D_{50}, D_{60} and D_{80} for all samples were determined by means of the fall tube test. For a limited number of samples (31) these values were again determined by means of the sieving method. In order to compare these values with other investigation results, the fall tube values were converted to sieve values. By means of the „smallest square method", the relations between fall tube values and sieve values of D_{10}, D_{50}, D_{60} and D_{80} were determined. Of all values obtained in this way the average D_{10}, D_{50}, D_{60} and D_{80} were determined for the whole dam. The results are given in Table 16. So the average uniformity coefficient D_{60}/D_{10} is 2.3.

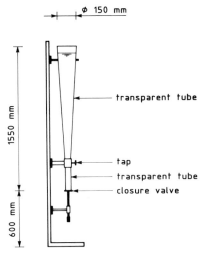

Fig. 125.   Arrangement of fall tube test.

Table 16.   Average fall tube values and sieve values.

magnitude	fall tube test (µm)	sieve test (converted) (µm)
D_{10}	225	197
D_{50}	436	387
D_{60}	520	448
D_{80}	834	685

The slope gradient and the average layer thickness could be determined by means of the cross-section profiles.

Other details:

– the thickness of hydraulically placed layers lies between 1 m and 3 m;

– the average achieved slope is 1 : 5 (from 1 : 4 to 1 : 7.5). This is defined as the slope between the crest line at approximately M.S.L. −5 m and the toe line at approximately M.S.L. −19 m.

21.6 Supplementary investigation

21.6.1 *In situ investigation*

The following tests were performed during and after hydraulic placing:

– 2 measurements of pore pressure;
– 13 Dutch cone tests after hydraulic placement;
– 1 electric density measurement after hydraulic placing;
– 1 nuclear density measurement after hydraulic placement;
– 2 borings after hydraulic placement.

The purpose of this investigation was to verify the predictions made during the design stage. One was especially interested whether the pore pressures and densities would develop in this type of sand as expected. Both data are required to carry out stability calculations according to the Bishop method.

In situ pore pressure measurements
From various investigations [2, 13] it appeared that the slope height, the specific sand production and the particle size have an essential influence on the slope gradient to be constructed. During the construction of the dam for the construction dock in the Europe basin, this was accounted for by an adapted hydraulic placement technique and selection of the reclaimed sand.
From these investigations it appeared that instability of a submerged slope can be traced by, among others, a sudden occurring increase of the pore pressure.
In order to check whether for the applied method of hydraulic placement such instability would also occur and at which slopes, two pore pressure measurements were performed. The pore pressure meters were placed in the sand bed in front and behind the spreader at an equal distance of approximately 15 m. Each pore pressure meter was placed in a sounding rod provided with a hoisting eye. After the probe was placed on the sand bed, it was further hammered into the sand by means of a drop weight. The situation at the bottom before starting the measurements, is indicated in Figure 126. The discharge height was approximately 3 m. During the test with the flattest slope gradient, no instability was observed at any time and also no excess pore pressure.
Subsequently, measurements were also carried out in a more landward position. In this position the difference in height with the first measuring position was approximately 2 m. A number of times instabilities were observed here by means of pore pressure variations.
In the Figures 127 and 128 some excess pore pressure measurements are presented. The highest pressures were approximately 13 kPa (more than 1 m water column). They

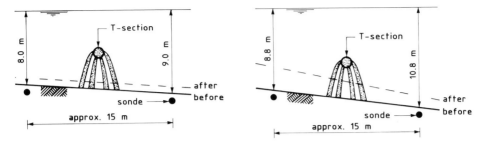

Fig. 126.   In situ pore pressure measurement.

occurred on the upper section of the slope. The duration was approximately 1 to 10 s. This was shorter than observed during the flow slides reported in [13] and [27], which can be explained with the presence of medium sand in combination with the relative small layer thickness of 2 m. The excess pressures at the toe were considerably smaller and amounted to only some centimetres water column. Sometimes a negative pressure was measured at the toe. The equilibrium slope after these instabilities was 1 : 7.5.

Figure 127 presents the situation during the passage of a tug. As a result of this an enormous fluctuation occurred of the pore pressure resulting in instabilities. During the first passage it was assumed that it was a matter of coincidence. After this phenomenon occurred again during the passage of the same tug, this was repeated on purpose for the third time. On all three occasions this resulted in instability.

The subsequent occurring disturbances could only be explained by the process of sand placement itself (see Fig. 128). The median particle size during the first phase was 600 µm. The equilibrium slope after each instability was approximately 1 : 7.5. After changing of a barge, of which the sand content had a diameter of 480 µm, no instabilities occurred for 15 minutes. The slope increased from 1 : 7.5 to 1 : 4.75. In the mean time the cover of the pore pressure meter positioned on the slope was 2.6 m and 1.4 m of the probe positioned at the toe. After reaching this gradient the slope became instable. The measurement was stopped because otherwise the probes would be lost.

From the measurements it appeared that the influence of accidental disturbances (such as in this case the passage of a tug) can have much greater influence than the excess pore pressure resulting from the hydraulic placement process.

Dutch cone tests and boring in the constructed dam

At the time of the geotechnical investigation during January and February 1989, the height of the dam varied between M.S.L. −6.1 m and M.S.L. −18.5 m. The cone resistance q_c appeared to vary between 2 MN/m² and 14 MN/m² up to a level of M.S.L. −20 m.

At M.S.L. −20 m a clay/peat layer was encountered with a q_c of 1 MN/m². Locally this clay/peat layer has been removed as a result of previous dredging. Below M.S.L. −20 m

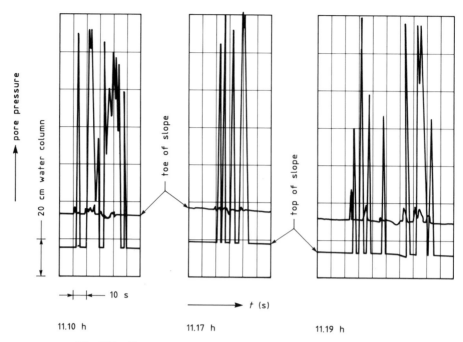

Fig. 127. Excess pore pressures as a result of the passage of a tug.

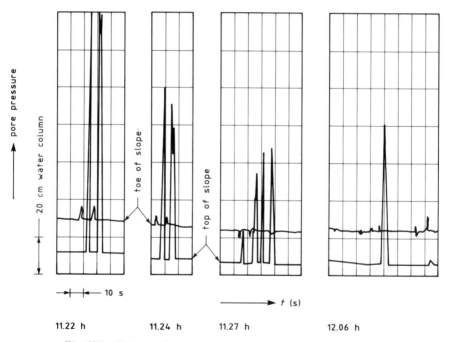

Fig. 128. Hydrostatic pressure as a result of the sand placing process.

232

up to M.S.L. -30 to -35 m, a predominantly sandy layer is encountered with q_c values of 4 to 20 MN/m². Here also silt and clay layers occur. Below M.S.L. -30 to -35 m the Pleistocene sand is encountered.

Density measurements in the dam
A nuclear and electric density measurement has been carried out. The results are mentioned in 21.7.1.

21.6.2 *Laboratory investigation*

In the laboratory the following tests were carried out:

– classification of the samples;
– determination of the densities and the water content;
– determination of the undrained shear strength on cohesive samples;
– determination of the grain size distribution;
– determination of the organic content;
– determination of the carbonate content;
– determination of the minimum and maximum dry densities;
– static triaxial tests for determining the angle of internal friction;
– determination of the critical density.

In Table 17 the results of the classification tests on sand are shown. The permeability coefficient k is mentioned as well, which was calculated with the formula of Hazen: $k = C \cdot D_{10}^2$ [64], in which k in m/day, D_{10} in mm and C mainly depends on the porosity. For these circumstance for C a value of 1000 is chosen.

Table 17. Results of classification tests.

depth rel. to M.S.L. (m)	D_{50} (mm)	D_{10} (mm)	$\dfrac{D_{60}}{D_{10}}$	k (m/day)	n_{max}*	n_{min}**
$-$ 8.5	0.18	0.11	2.0	12	0.48	0.36
-10.5	0.51	0.25	3.0	63	0.42	0.31
-11.5	0.68	0.24	4.8	58	0.40	0.32
-13.3	0.35	0.13	3.1	17	0.43	0.30
-13.8	0.27	0.11	3.4	12	0.40	0.32
-14.0					0.45	0.32
-14.5	0.21	0.11	2.5	12	0.40	0.36
-15.5	0.27	0.13	3.4	17	0.39	0.35
-17.0					0.44	0.31
-17.5	0.31	0.14	2.4	20	0.42	0.30
-20.35	0.16	0.07	2.5	5	0.48	0.37
-21.0					0.50	0.36
-23.0					0.48	0.35

* n_{max} = maximum porosity of dry soil
** n_{min} = minimum pososity of dry soil

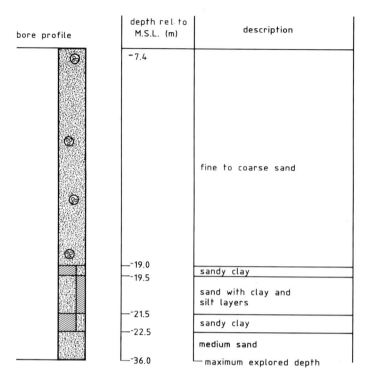

Fig. 129. Bore profile of underwater dam.

Triaxial tests were performed on two samples. Also 5 critical density tests were performed according to the so called „wet method". For this purpose a sand sample was prepared in the triaxial cell at a selected density (porosity) and saturated with water. Subsequently the sample was brought to failure relative rapidly under undrained conditions. From the measured force, deformation and hydrostatic pressure the critical density can then be determined (see 11.6.6).

The results of these tests are shown in Table 18 and Figure 129 gives a typical bore profile in the underwater dam.

Table 18. Results of critical density tests.

depth rel. to M.S.L. (m)	n_{max}	n_{min}	n_{crit}
− 8.5	0.48	0.36	0.46
−10.5	0.42	0.31	0.40
−11.5	0.40	0.32	0.41
−13.3	0.43	0.30	0.42
−17.5	0.42	0.30	0.40

234

21.7 Evaluation

21.7.1 *Discussion of the investigation results*

General

In the following section the investigation results (see 21.6) are compared with the advised procedures. Wherever possible, the agreements and the deviations are explained and clarified. This comparison will be related to the following subjects:

- method of hydraulic placement;
- requirements for sand;
- slope gradient;
- density of the fill;
- development of hydrostatic pressure during hydraulic placement.

Method of hydraulic placement

The dam was constructed with sand for which a number of possibilities could be considered for the method of sand placement, namely:

- hydraulic discharging;
- lump-like discharging;
- curtain-like discharging.

Because of the differences between these placing methods, different porosities result. It appears that in general for hydraulic discharging higher porosities result than for other methods.

From field measurements it appears that for hydraulically placed sand, underwater slopes of 1:5 can be achieved.

Except for the method of placement, also the grain size of the sand is of influence for the slope to be made.

With respect to the work method to be applied it was advised at the time (see 21.2.1) to apply vertical hydraulic discharging or spraying. The dam was subsequently build up by means of vertical spraying.

Requirements for sand

In order to build the dam as steep as possible, conditions were defined for the grain size distribution of the sand to be applied. In particular the uniformity coefficient D_{60}/D_{10} and the D_{10} are of importance because of the compactability and the permeability of the sand.

It was advised to use a type of sand with a D_{60}/D_{10} of 2.5 and a D_{10} greater than 150 μm. Of the sand used, D_{60}/D_{10} varied between 2 and 4.8, whereas D_{50} varied between 160 μm and 680 μm and D_{10} between 70 μm and 250 μm. This means that the average grain size of the used sand is coarser than the minimum required conditions. This entails that a steeper slope is possible and that the excess hydrostatic pressures as a result of the construction, will not occur or will occur to a lesser extent.

235

Slope gradient

On the basis of the performed analyses an obtainable gradient was advised of $1:6$ to $1:7$. This advice was also based on the minimum sand requirements.

For the realization of the underwater dam it appeared that a gradient of $1:5$ was possible. This is the result of using a considerably coarser sand and a layer by layer hyraulic placement by spraying in a vertical direction. Because of a greater risk of collapse a slope of $1:4$ was not build on purpose.

Density of the fill

The density with which the sand builds up during placing, can be estimated with the formula:

$$n_{dam} = 1 - 0.83(1 - n_{min})$$

in which:

$$n_{dam} = \text{porosity in the dam}$$
$$n_{min} = \text{minimum porosity}$$

This formula is advised by HEEZEN and VAN DER STAP [2] and is based on tests by BAGNOLD on uniform sand. With this formula the in situ porosities are predicted based on n_{min}.

Based on an expected maximum density or minimum porosity of $n_{min} = 0.37$ to 0.38 it follows that the sand has a dry density of 1350 to 1450 kg/m^3 ($n_{dam} = 0.48$ to 0.49). With measurements (both in situ and on samples) porosities can be determined. The results of predictions and measurements are summarized in Table 19.

Table 19. Results of density measurements and tests on samples of borings in the constructed dam.

depth relative to M.S.L. (m)	n_{min}	n in situ				n from samples
		predicted n_{dam}	measured			
			electric		nuclear	
			method A	method B		
− 8.5	0.36	0.47	–	–	–	0.46
−10.5	0.31	0.43	0.45	0.41	–	0.46
−11.5	0.32	0.44	0.41	0.36	0.34	0.37
−13.3	0.30	0.42	0.40	0.38	0.38	0.42
−17.5	0.30	0.42	0.40	0.38	0.36	0.42

From this Table it appears that the values measured on the samples and the electric density measurement according to method A (greater distance between the electrodes hence large measured volume, minor influence of compaction), correspond reasonably well with the predicted values when based on the data of the used sand.

236

The nuclear density measurement provides the lowest n values. This is caused by the relatively large measuring device and the small volume measured.

The prediction for the expected porosity based on the earlier investigation, agrees rather well for the electric density measurement as well as for the samples. The relatively large deviation with the nuclear density measurement and with the electric density measurement B (with a small distance between the electrodes), is caused by the local compaction as a result of inserting the probe and because the measured volume is much smaller compared with the electric probe according to method A. This phenomena, measuring too high densities with the nuclear probe, especially occurs in loose sand formations. This measuring technique will be better applicable in dense sand (compare 11.6.2 and 11.6.3).

Development of hydrostatic pressure during hydraulic placement

During hydraulic placement two hydrostatic measurements were performed. The results are presented in the Figures 127 and 128. From these measurements it appears that in the upper 0.5 to 1.0 m, negative hydrostatic pressures as well as positive hydrostatic pressures develop. The magnitude of these excess hydrostatic pressures was approximately 10 to 15 kPa.

This means that the sand liquefies and will flow as a result. Presumably small flow slides occurred regularly during the process of hydraulic placement. The reason that this did not result in flatter slopes, probably depended also on the great permeability and the thin layers: excess pore water is already squeezed out after 1 to 10 *s*, whereby the liquefied sand has no time to flow into a flatter slope.

21.7.2 *Conclusions*

- Through placement in thin layers, 1 to 3 m, the occurrence of large excess pore pressures (more than approximately 10 kPa) was prevented while steep slopes were achieved.
- A slope of 1 : 5 could be constructed through pipeline discharging with the pipe just above the sand, at which the sand-water mixture was spread across a large width (spraying).
- From the study it appeared that the underwater dam was stable at a slope steeper (1 : 5), than originally advised (1 : 6 to 1 : 7). This was presumably possible because considerably coarser sand was used with little silt and clay, also because silt and clay was washed overboard during barge loading.
- By defining minimum requirements for sand with respect to grain size distribution and minimum grain diameter, a sound product can be delivered if at the same time also a careful working method is maintained.
- The electric density measurement according to method A provides the most realistic picture of the sand density. It must be said that the measurement rapidly becomes unreliable if many thin silt and clay layers occur.

- The nuclear density measurement results in densities which are too high due to the large effect of displacement by the probe itself.
- It appears that with the formula mentioned in 21.7.1, the expected density after hydraulic placement can be calculated with reasonable accuracy.
- In this example the stability was dependent on the soil mechanical aspects such as the presence of a slope, the triggering of instability by passing ships and the steeper building up of the slope as a result of sedimentation.
- By carrying out a soil investigation in advance in the expected borrow area, the composition of the sand can be investigated and design calculations can be performed on the basis of these data in relation with the working method to be selected.

CHAPTER 22

SUMMARY

In this manual the theory and the practice of artificially constructed sand fills in water is presented. On the basis of thirteen case histories the design practice is highlighted. By means of detailed elaboration of a practical case the theory is again more extensively verified with project experience.

The theoretical part of this manual provides a description of the process phenomena for the construction with sand in water based on theory development of the physical processes supported by observations of these processes in the field and observations in the laboratory. With respect to the placement of sand in water a distinction can be made between placement above water and placement under water.

The dimensions of a sand fill placed in water are determined by the dimensions of the crater developed during this process, the slope and the sedimentation characteristics of the sand-water mixture.

For a proper understanding of the formation of a sand fill the following phenomena are most important:

- the behaviour of a sand-water mixture after leaving the discharge pipe or hopper;
- crater formation as a result of the impinging on the bottom of a sand-water mixture;
- the formation of respectively the above and under water fill area;
- the occurrence of flow slides.

If a sand-water mixture flow rate and concentration are known at the point of discharge from a pipeline or hopper vessel, then, for a given shape and dimension of discharge opening and for given field conditions the following parameters can be predicted:

- the dimensions;
- the concentrations and velocities of the mixture along the trajectory;
- the underwater trajectory of the sand-water mixture.

Subsequently from these characteristic parameters the dimensions of the crater, which develops when the jet impinges on the bottom, can also be predicted.

Besides the grain diameter, for the determination of the slope gradient, the specific flow rate is the most important parameter for the abovewater fill area and the specific sand production rate for the underwater fill area. A large specific flow rate and a large specific sand production rate result in flatter slopes.

On the abovewater fill area either channel formation may develop or systems of cascades, hydraulic jumps and terraces. These bed formations are dependent on the size of the specific flow rates. For large specific flow rates the channel formation dominates.

When distributing the mixture flow across the fill area small specific flow rates are obtained and systems of cascades, hydraulic jumps and terraces develop together with relatively steep slopes.

In an underwater fill area in principle the same bed formations may occur as on the abovewater fill area; i.e. channel formation or systems of cascades, hydraulic jumps and terraces. Sedimentation is determined by the size of the specific sand production rate and the dimensions of the grains. The final slope develops as a result of a combination of sedimentation and flow slides.

The occurrence of flow slides mainly depends on the sedimentation length, the porosity of the deposited sand and on the height of the initial deposited slope. On the basis of related parameters the possible occurrence of flow slides when placing sand is clarified. In this manual a method is described for the determination of the distribution width of a sand-water mixture. From this, the specific flow rate is derived from which the slope gradient is determined.

Apart from an analyses of the possible occurrence of flow slides, which covers also existing slopes, measures are discussed to limit the chance on flow slide occurrence. At the same time a number of methods are discussed to measure the in situ density of a volume of sand.

The theoretical section of the manual concludes with a summary of the theory for practical use.

In the practical section, an enumeration is given of the different types of structure for which hydraulic placing of sand in water may be applied as well as the functions of that sand within these structures. The additional phenomena which accompany the application of sand and the subsequent design and construction requirements are discussed. With respect to the construction method an enumeration is given of the most important available types of equipment together with the respective achievable tolerances and workability restrictions. Attention is also paid to the measurement of the produced quantities. A chapter is devoted to quality assurance of design and construction.

In a number of case histories some frequently occurring applications of hydraulic placement of sand are discussed. A number of typical cases are matched with the discussed theory.

A case history in which the use of the theory is applied step by step is examined in more detail, providing a detailed insight into the practical application of the theory.

Finally it should be stated that, with the theory presented in this manual, a reliable description of the phenomena is provided in a qualitative sense. Most of the time, a sufficiently quantitative approach is not possible, which means that the accuracy of calculation results is often not more than a factor 1.5 to 2. In future this accuracy may be improved with among others the results of the questionaire.

APPENDIX A

DISCHARGE PROCESS PARAMETERS

A1 Introduction

In this Appendix the parameters, which are used in this manual for the description of the discharge process, are defined in more detail. It is assumed herewith that the reader of this manual is acquainted with basic concepts such as the in situ soil density (wet or dry), the particle density, the porosity and the void ratio. These basic concepts form the materials for the discharge process parameters described hereafter. In this Appendix only some characteristic values of these basic concepts are included in a Table. The discharge process parameters can be sub-divided in various categories, namely parameters related to the production, parameters related to the behaviour of the mixture on the fill area and parameters concerning the volume change of the dredged soil during the different phases of the dredging process. Indications will be given of practical values for these parameters.

The parameters related to the production are:

- mixture density;
- mixture flow rate;
- concentration.

The concentration is sub-divided in:

- volume concentration (=particle concentration);
- transport concentration;
- apparent concentration.

The parameters in relation to the behaviour of the mixture on the fill area are:

- specific mixture flow rate;
- specific sand production rate;
- specific sand transport rate.

These parameters are less common in the dredging practice, but are used for the theoretical description of the process. For this reason, these are discussed in more detail here.

Note: Also the mixture density, the mixture flow rate and the concentration are of importance on the fill area.

The parameter in relation to the change of volume of the dredged soil is:

- bulking.

A2 Density, porosity and void ratio of soil

In Table A1 an indication is given of some characteristic values of the density, the porosity and the void ratio for some types of soil. In practice the composition of the soil varies considerably and these parameters have to be determined for each case with specific tests.

Table A1. Practical values for the density, the porosity and the void ratio for some types of soil.

soil type	wet density ϱ_n (kg/m³)	dry density ϱ_d (kg/m³)	porosity n (%)	void ratio e
peat	<1000–1100	100– 300	60–85	1.5–5.6
silt	1200–1300	300– 500	80–90	4 –9
soft clay	1400–1600	500–1100	60–80	1.5–4
firm clay	1800–2000	1300–1700	35–50	0.5–1
clayey sand	1800–2000	1300–1600	40–50	0.7–1
loose sand	1700–1900	1200–1500	45–55	0.8–1.2
dense sand	1900–>2000	1500–1700	35–45	0.5–0.8

Depending on the composition of the grains the particle density varies between 2600 kg/m³ and 2700 kg/m³.

A3 Parameters in relation to the dredging process

A3.1 Mixture density

The mixture density is the density of the sand-water mixture ϱ_m in the discharge pipeline. Its value depends on many factors such as, among others, the composition of the soil to be dredged, the pipeline diameter, the discharge distance, the installed discharge capacity, the diameter of the grains to be transported and not in the least the skill of the dredging crew. The value of the mixture density varies roughly between 1100 kg/m³ and 1400 kg/m³, and can even be more than 1500 kg/m³ under favorable circumstances.

A3.2 Mixture flow rate and mixture velocity

The flow rate Q of the sand-water mixture through the pipeline, is defined as the mixture volume transported through a cross-section per unit of time. The mixture velocity is defined as follows:

$$v_m = \frac{Q}{A} \qquad (m/s)$$

in which:

Q = mixture flow rate (m³/s)
v_m = velocity of the sand-water mixture in the pipeline (m/s)
A = cross-section of the pipeline (m²)

242

Also the velocity in the pipeline depends on many factors such as again the pipeline diameter, the discharge distance, the installed discharge capacity, the diameter of the grains to be transported and also the density of the mixture in the pipeline and the working conditions. In practice this means that this value can vary considerably: between approximately 3 to 9 m/s.

Needless to say that the flow rate also depends on the above mentioned factors and may vary as well for different situations. Values between 0.5 m^3/s and 4 m^3/s do occur in practise.

A3.3   Concentration

Volume concentration

The volume concentration (or the percentage of solids) is the ratio between the volume of the grains and the total volume of the mixture at a certain place and at a certain moment:

$$c_v = \frac{V_s}{V_m}$$

in which:

c_v = volume concentration
V_s = volume of the solids (m^3)
V_m = volume of the total mixture (m^3)

Per unit of volume the following relation applies:

$$\varrho_s c_v + \varrho_w(1 - c_v) = \varrho_m$$

or:

$$\varrho_w(1 + \varDelta c_v) = \varrho_m$$

in which:

$$\varDelta = \frac{\varrho_s - \varrho_w}{\varrho_w}$$

From this follows the well known relation:

$$c_v = \frac{\varrho_m - \varrho_w}{\varrho_s - \varrho_w} \tag{A1}$$

in which:

ϱ_m = mixture density (kg/m^3)
ϱ_w = water density (kg/m^3)
ϱ_s = particle density (solids) (kg/m^3)

In practice the volume concentration can be measured on board of a dredger. The measuring instruments generally contain the relation between the densities and the concentration and must be calibrated for applicable values. The mixture density can directly be read. If for example the concentration amounts to 17 % and the density of the water is 1025 kg/m³ and of the particles 2650 kg/m³, than the mixture density equals 1300 kg/m³.

Transport concentration

The concept of transport concentration applies to the production of the solids of a sand-water mixture and equals the ratio between the solids production and the total production of the sand-water mixture:

$$c_t = \frac{Q_s}{Q_m}$$

in which:

c_t = transport concentration
Q_s = solids production (flow rate) (m³/s)
Q_m = mixture production (flow rate) (m³/s)

Just like the volume concentration this definition concerns only the solids and not the pores. In this definition of concentration the respective velocities of the solids and the fluid are involved.

For the production of the solids the following applies:

$$Q_s = \tfrac{1}{4}\pi D^2 v_s c_v \qquad\qquad (\text{m}^3/\text{s})$$

and for the production of the mixture:

$$Q_m = \tfrac{1}{4}\pi D^2 v_m \qquad\qquad (\text{m}^3/\text{s})$$

In this formulae:

v_s = velocity of the solids (m/s)
v_m = velocity of the mixture (m/s)
D = pipeline diameter (m)

For the relation between the transport concentration and the volume concentration now the following applies:

$$\frac{c_t}{c_v} = \frac{v_s}{v_m} = f_t$$

or:

$$c_t = \frac{\varrho_m - \varrho_w}{\varrho_s - \varrho_w} f_t \qquad\qquad (\text{A2})$$

The factor f_t is defined as the transport factor. Because the velocity of the solids is always somewhat less than the velocity of the mixture, the value of f_t is always smaller than 1.

Apparent concentration

The apparent concentration owes its name to the fact that the solids as well as the pores are included in the definition. This concept applies to the sand production on the fill area. In order to determine its value the porosity on the fill area must be known. In practice this value is often estimated in first instance. For the sand production the following relation applies:

$$Q_g = \frac{Q_s}{1-n}$$

in which:

Q_g = sand production (including pores) (m^3/s)
Q_s = solids production (without pores) (m^3/s)
n = porosity

For the apparent concentration applies:

$$c_g = \frac{Q_g}{Q_m} = \frac{Q_s}{(1-n)Q_m}$$

or:

$$c_g = \frac{c_t}{1-n}$$

After substitution of c_t with formula (A2) it follows for the relation between the apparent concentration and the densities, that:

$$c_g = \frac{\varrho_m - \varrho_w}{\varrho_s = \varrho_w} f_t \frac{1}{1-n}$$

For homogeneous transport where f_t equals approximately 1, the apparent concentration can be defined with the well known formula:

$$c_g = \frac{\varrho_m - \varrho_w}{\varrho_s - \varrho_w} \cdot \frac{1}{1-n} \tag{A3}$$

Based on the typical values for the porosity of 0.4, the density of water of 1.0 t/m^3 and the particle density of 2.65 t/m^3, this results in the following rule of thumb in the dredging practice:

$$c_g \approx \varrho_m - 1$$

If for example the density of the mixture amounts to 1.35 t/m^3, it follows from this formula that the apparent concentration equals approximately 0.35. With this concentration, the produced quantity of sand on the fill area can roughly be determined.

A4 Parameters in relation to the behaviour of the mixture on the fill area

As already mentioned in the introduction, the parameters discussed below are mainly used for the theoretical description of the discharge process on the fill area.

A4.1 Specific mixture flow rate

The specific mixture flow rate is the mixture flow rate on the fill area per unit width. In practice it is generally difficult to determine this width. If the sand-water mixture flows through a channel and deposits at the water line, the discharge width equals the width of the channel. For an even distribution of the sand-water mixture between the bunds, the discharge width equals the distance between the bunds. This is for instance the case with sand closures where one is primarily interested in a small crest in order to realize a closure with a minimum quantity of sand. For the specific flow rate it applies that:

$$q = \frac{Q}{b}$$
(A4)

in which:

q = specific flow rate (m²/s)
Q = total flow rate (m³/s)
b = discharge width (m)

E.g. for a sand production rate of 2.5 m³/s and a discharge width varying between 30 m and 50 m, the mixture flow rate amounts to 0.05 to 0.08 m²/s.
E.g. in the event of channel formation on the fill area, the mixture flow rate in the channel amounts to approximately 0.6 m²/s for a channel width of 4 m.

A4.2 Specific sand production

The specific sand production is the sand production on the fill area per unit of discharge width. The sand production is calculated from the mixture flow rate and the apparent concentration, therefore:

$$P = c_g Q \qquad\qquad (\text{m}^3/\text{s})$$

in which:

P = sand production (m³/s)
c_g = apparent concentration
Q = mixture flow rate (m³/s)

The specific sand production is now equal to:

$$p = \frac{P}{b} \qquad\qquad (\text{m}^2/\text{s}) \qquad\qquad (\text{A5})$$

A4.3 Specific sand transport

In the dredging practice the produced quantity of sand (including pores) is often expressed in m³, thus in volume. However, for theoretical purposes, the production is preferably expressed in kg weight of the solids for the sake of uniformity. The specific sand transport is the sand production per unit of width expressed in kg solids. In formula:

$$s = c_v \varrho_s q$$

or:

$$s = \frac{P \varrho_s (1 - n)}{b} \qquad\qquad \text{(kg/ms)} \qquad\qquad \text{(A6)}$$

in which:

$\quad s$ = specific sand transport (kg/ms)
$\quad P$ = sand production (m^3/s)
$\quad b$ = discharge width (m)
$\quad n$ = sand porosity on the fill area
$\quad \varrho_s$ = particle density (solids) (kg/m^3)

E.g. for a porosity n of 0.4, a particle density ϱ_s of 2650 kg/m^3, a mixture flow rate of 2.5 kg/m^3 and a concentration of 0.35, the specific sand transport amounts to respectively 27 kg/ms and 46 kg/ms for a discharge width of 50 m and 30 m.

A5 Parameters in relation to the change of volume of the dredged soil

A5.1 *Bulking and debulking*

When handling soil mechanical or hydraulically, change of volume occurs. This is called bulking or debulking. Change of the soil volume in relation to the dredged volume can occur in a hopper dredger, in a barge or on the fill area. Difference in volume is also encountered between the volume in the bin and the volume on the discharge area.

With hydraulic transport change in volume is caused by the sedimentation process. Sand in the bin or on the fill area generally is less compacted than sand encountered in the borrow area. Therefore densities in the bin and on the fill area are less than the in situ density. This mainly applies to sand placed under water. On the fill area, in the tidal zone and above water, sand generally is very well compacted because of the action of bulldozers.

Lower densities than the in situ densities result in volume increase. However, this volume increase may be compensated by the loss of the fine fraction due to overflowing of a hopper dredger or when loading barges and with the outflowing of process water at the down stream end of an abovewater fill area. This change in volume is also dependant on the particle distribution. For a small portion of the fines in the dredged sand, the disappeared fine fraction may be replaced by an increased pore volume, in that case no change of the soil volume will take place.

Higher densities than the in situ density result in volume decrease.

Causes for change in volume during the dredging process can be summarized as follows:

- bulking or debulking;
- loss of fines during overflowing;
- loss of fines on the fill area.

The total change of volume is often a combination of these factors and varies from project to project.

When determining the required sand quantity for reclamation of a certain area or placement of a sand body in water, for which the change of volume as a result of the dredging process has to be known, also consolidation of the soil has to be allowed for in addition to the factors mentioned above.

In this paragraph only the phenomenon of bulking and debulking will be discussed. The bulking or debulking is defined by:

$$B = \frac{\text{volume after handling}}{\text{volume before handling}}$$

The bulking (or debulking) concept is illustrated in Figure A1, in which a volume increase is indicated. In this Figure:

V_1 = soil volume before bulking (m³)
V_2 = soil volume after bulking (m³)
n_1 = porosity before bulking
n_2 = porosity after bulking

In Figure A1 the pore volume increases, whereas the volume of the solids remains the same, therefore:

$$(1 - n_1)V_1 = (1 - n_2)V_2$$

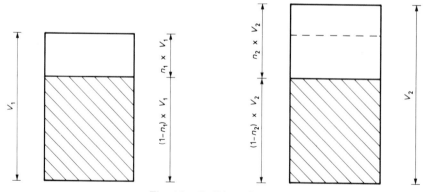

Fig. A1. Bulking of soil.

Per definition:

$$B = \frac{V_2}{V_1}$$

From this follows:

$$B = \frac{1 - n_1}{1 - n_2} \qquad\qquad\qquad\qquad (A7)$$

Because for completely saturated soil the following formula applies for the porosity:

$$n = \frac{\varrho_s - \varrho_g}{\varrho_s - \varrho_w}$$

the following general formula can be inferred for bulking:

$$B = \frac{\varrho_1 - \varrho_w}{\varrho_2 - \varrho_w} \qquad\qquad\qquad\qquad (A8)$$

In this formula:

$\varrho_1 = $ density of the soil before bulking (kg/m^3)
$\varrho_2 = $ density of the soil after bulking (kg/m^3)
$\varrho_w = $ water density (kg/m^3)
$\varrho_s = $ particle density (solids) (kg/m^3)

With formula (A7) bulking as a result of soil handling can be calculated from the porosities. When before handling the porosity of the soil equals for instance 0.4 and after handling 0.45, then the bulking factor amounts to 1.09. Hence the total soil volume increases with 9 % and not with 5 % as is often wrongly concluded on the basis of porosities.

With formula (A8) bulking as a result of soil handling can be assessed on the basis of densities. When the density of the soil in the borrow area equals for example 2000 kg/m^3 and the density in the fill area is 1950 kg/m^3, than the bulking factor amounts to 1.05. As a result of density decrease, the volume in the fill area is therefore 5 % larger than the corresponding volume in the borrow area.

Bulking is especially important if silty or clayey material has to be dredged. This material has a considerable lower density in the fill area than in situ. In that case the bulking may be 130 % to 140 %.

APPENDIX B

DRILLING ISLANDS ISSUNGNAK AND UVILUK

B1   General

For the construction of an artificial island such as Issungnak, the volume of sand can be minimized in two ways: by realizing the smallest possible beach profile and by making the steepest possible underwater slopes. The shape of the beach profile as a function of grain and wave characteristics can be described by means of [58] and is outside the scope of this manual. However, with the theory presented in this manual an estimate can be made of the underwater slopes as a function of the grain characteristics and working method. This can also be done for the island Uviluk.

In the following the results of some calculations are presented in which the influence on the underwater slope of the applied working method will be demonstrated. Three construction methods are involved for Issungnak:

- bottom discharging with split barges;
- pipeline discharging from the water surface using a stationary suction dredger;
- pipeline discharging on an abovewater fill area using a stationary suction dredger.

For Uviluk one construction method is involved, namely:

- pipeline discharging via spreader just above the s from a hopper dredger.

B2   Bottom discharging with split barges (Issungnak)

For the bottom discharging with split barges it is assumed that the bin length is 30 m and the bin volume equals 30 m × 12 m, the bin opens with a velocity of approximately 0.1 m/s and is emptied within about 20 s. It is furthermore assumed that the sand-water mixture leaves the hopper with a concentration c_0 of approximately 0.5. The maximum width of the opening during discharging is approximately 1 m. For an average width of the opening b_0 of approximately 0.6 m the velocity U_0 with which the mixture leaves the hopper will be equal to approximately 1 m/s. It is further assumed that the sand has the following characteristics: $D_{50} = 225$ μm and $D_{15} = 150$ μm. The estimated rather high value for D_{15} is also based on the assumption that possible occurring silt fractions largely flow overboard during barge loading.

In correspondence with the concepts presented in 4.1.1 it is assumed that bottom discharging with split barges can be considered to be curtain like discharging. The calculations of the concentration c_b, the velocity U_b and the jet width b_b with which the sand-water mixture hits the sea bed, are carried out with an adapted version for curtain like discharging of the computer program „STRAAL3D" (see 4.2). The results will be sufficiently reliable to obtain a proper impression of the falling process.

250

It is assumed that the falling height beneath the hopper opening varies from 17 m (original sea bed) to 3 m. It is further assumed that the current velocity U_a is always limited to 0.5 m/s. The horizontal displacement of the centre of the jet is indicated with x_b. The calculations are carried out for a minimum opening width of 0.4 m and a maximum opening width of 1.0 m.

Some calculation results are presented in Table B1.

Table B1. Falling of sand through water when discharging from a split barge.

U_0 (m/s)	b_0 (m)	U_a (m/s)	fall height 17 m				fall height 10 m				fall height 3 m			
			c_b	U_b (m/s)	b_b (m)	x_b (m)	c_b	U_b (m/s)	b_b (m)	x_b (m)	c_b	U_b (m/s)	b_b (m)	x_b (m)
0.5	0.4	0	0.02	2.7	1.15	0	0.053	3.2	0.68	0	0.2	2.6	0.19	0
1.0	0.4	0	0.03	3.3	1.20	0	0.078	3.9	0.72	0	0.25	3.2	0.25	0
1.5	0.4	0	0.04	3.8	1.20	0	0.094	4.4	0.75	0	0.3	3.7	0.27	0
0.5	1.0	0	0.08	5.0	1.05	0	0.23	5.6	0.62	0	0.25	4.3	0.23	0
1.0	1.0	0	0.11	6.0	1.20	0	0.27	6.4	0.72	0	0.3	4.7	0.35	0
1.5	1.0	0	0.13	6.7	1.25	0	0.30	6.8	0.85	0	0.38	5.0	0.40	0
0.5	0.4	0.5	0.01	1.25	2.0	11	0.031	2.0	1.0	3.8	0.2	2.6	0.19	0.1
0.5	1.0	0.5	0.07	4.3	1.2	2.5	0.22	5.3	0.65	0.6	0.25	4.3	0.23	0.0

For the extreme values of this table the characteristic parameters of the sand discharge process under water are calculated.

Because here it concerns plain shaped jets, the crater depth resulting from a clear water jet is calculated using formula (5) of RAJARATNAM (see 5.1.1). This depth is reduced for the influence of sand by means of Figure 41 (see 5.1.2). The crater diameter $2r_\infty$ is found by means of formula (6) (see 5.1.2). Where applicable, the crater depth is subsequently reduced up to the maximum which follows from the consideration that the crater volume cannot be greater than the volume of the inflowing mixture (=volume of split barge times c_0/c_b) (see 5.2).

The maximum width B_{max} of the mixture flowing over the crater edge is supposed to be equal to twice the length of the split barge plus twice the crater width: $B_{max} = 2 \cdot 30 + 2 \cdot 2r_\infty$. The width B_{min} is found by means of the Figures 44 and 45 (see 5.3). Because it is likely that the sand-water mixture flows out of the crater via channels, B_{min} will further be used in the calculations. The values of the specific flow rate q and the specific sand production rate s are subsequently calculated with the formulae (13) and (14) respectively. The sedimentation length is found by extrapolation of Figure 48 (see 6.3). This is in fact not correct because the mixture flow is not present for longer than approximately 20 s whereas the sedimentation period amounts to some hundreds of seconds (see note 1 in 6.3). The actual sedimentation length will be considerably shorter. Finally the slope gradient is determined from Figure 56a (see 8.5). The results are summarized in Table B2.

Table B2. Crater dimensions and sedimentation length when discharging from a split barge.

magnitude	unit	fall height 17 m	fall height 10 m	fall height 3 m
crater width $2r_\infty$	m	15– 40	10– 30	5– 20
B_{max}	m	90–140	80–120	70–100
B_{min}	m	25– 40	20– 30	6– 15
specific flow rate q	m²/s	3– 7	3– 6	1– 3
specific sand production s	kg/ms	80–2,300	250–4,500	1,300–4,000
sedimentation length L	m	300–700	300–600	200–600
equilibrium slope $\tan \alpha$	–	$1:8$–$1:30$	$1:13$–$1:40$	$1:20$–$1:40$

On the basis of average values for the equilibrium slope given in Table B1, the manner in which the slope could have been built up is indicated in Figure B1 and the gradient of the resulting slope if no flow slides occurred. According to the calculations this slope could not have been much steeper than $1:20$, even when the split barge is opened slowly, because the minimum fall height (3 m) is critical (see Fig. B1). Moreover it appears from this Figure that the lower part of the slope must become flatter as a result of the large sedimentation length. fIn reality the achieved slope was steeper in the upper reach, namely $1:10$. Such a slope does follow from the calculations if it is assumed that the mixture is spread across a much larger width than B_{min}, for example $B = B_{max}$.

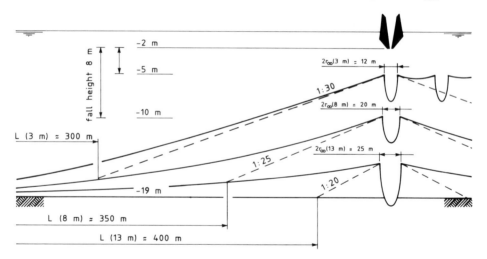

Fig. B1. Building up of the island Issungnak using split barges.

B3 Pipeline discharging from the water surface (Issungnak)

For pipeline placement from the water surface it is supposed that the diameter d_0 of the discharge pipeline of the stationary suction dredger was 0.7 to 0.9 m, the concentration c_0 at the point of discharge 0.14 to 0.20 and the mixture velocity U_0 4 to 5 m/s. It is further assumed that the sand had the following characteristics: $D_{50} = 225$ μm and

252

$D_{15} = 110\ \mu\text{m}$ (hence no washing out of fines). It is also assumed that the pipeline nozzle was kept approximately 1.5 m above the water surface at all times, sometimes in a horizontal position, sometimes in a vertical position. Characteristic parameters of the sand water mixture close to the sea bed are again calculated with the computer program „STRAAL3D". Some results are presented in the Tables B3 to B6.

Table B3. Falling of sand through water for pipeline discharging in horizontal direction, $d_0 = 0.7$ m.

				fall height 19 m				fall height 11 m				fall height 3 m		
U_0 (m/s)	c_0 (m)	U_a (m/s)	c_b	U_b (m/s)	d_b (m)	x_b (m)	c_b	U_b (m/s)	d_b (m)	x_b (m)	c_b	U_b (m/s)	d_b (m)	x_b (m)
4	0.14	0	0.016	2.8	3.4	9.7	0.032	3.3	2.2	7.9	0.12	6.1	0.86	4.2
5	0.14	0	0.018	3.0	3.6	12	0.033	2.4	2.4	10	0.12	6.7	0.90	5.5
4	0.20	0	0.023	3.2	3.2	9.1	0.045	3.7	2.2	7.6	0.17	6.2	0.85	4.5
5	0.20	0	0.024	3.4	3.4	11	0.047	3.9	2.3	9.6	0.18	6.7	0.90	5.5
4	0.14	+0.5	0.008	1.6	5.7	15	0.017	2.1	3.4	9.8	0.072	4.1	1.25	4.4
4	0.14	−0.5	0.011	2.0	4.1	2.5	0.023	2.7	2.6	4.5	0.070	4.0	1.3	4.1

Table B4. Falling of sand through water for pipeline discharging in horizontal direction, $d_0 = 0.9$ m.

				fall height 19 m				fall height 11 m				fall height 3 m		
U_0 (m/s)	c_0 (m)	U_a (m/s)	c_b	U_b (m/s)	d_b (m)	x_b (m)	c_b	U_b (m/s)	d_b (m)	x_b (m)	c_b	U_b (m/s)	d_b (m)	x_b (m)
4	0.14	0	0.022	3.3	3.5	10	0.042	4.1	2.4	8.4	0.14	6.8	0.95	4.7
5	0.14	0	0.024	3.5	3.7	13	0.044	3.9	2.5	11	0.14	7.4	1.04	5.9
4	0.20	0	0.030	3.8	3.4	9.8	0.060	4.5	2.3	8.1	0.20	6.8	0.95	4.7
5	0.20	0	0.032	4.0	3.6	12	0.060	4.4	2.4	10	0.20	7.4	1.04	5.9
4	0.14	+0.5	0.018	2.8	4.0	5.2	0.033	3.3	2.8	6.0	0.11	5.4	1.30	4.4
4	0.14	−0.5	0.013	2.3	3.2	13	0.028	2.9	3.2	9.5	0.11	5.4	1.25	9.4

Table B5. Falling of sand through water for pipeline discharging in vertical direction, $d_0 = 0.7$ m.

				fall height 19 m				fall height 11 m				fall height 3 m		
U_0 (m/s)	c_0 (m)	U_a (m/s)	c_b	U_b (m/s)	d_b (m)	x_b (m)	c_b	U_b (m/s)	d_b (m)	x_b (m)	c_b	U_b (m/s)	d_b (m)	x_b (m)
4	0.14	0	0.020	3.3	3.4	0	0.040	4.1	2.2	0	0.14	4.5	0.66	0
5	0.14	0	0.022	3.6	3.4	0	0.043	4.6	2.2	0	0.14			0
4	0.20	0	0.028	3.7	3.3	0	0.055	4.5	2.1	0	0.20			0
5	0.20	0	0.031	4.0	3.4	0	0.060	5.0	2.2	0	0.20	5.5	0.67	0
4	0.14	0.5	0.013	2.4	4.3	4.2	0.028	3.2	2.6	1.5	0.14			0

253

Tabel B6. Falling of sand through water for pipeline discharging in vertical direction, $d_0=0.9$ m.

U_0 (m/s)	c_0 (m)	U_a (m/s)	c_b	U_b (m/s)	d_b (m)	x_b (m)	c_b	U_b (m/s)	d_b (m)	x_b (m)	c_b	U_b (m/s)	d_b (m)	x_b (m)
			fall height 19 m				fall height 11 m				fall height 3 m			
4	0.14	0	0.027	3.9	3.4	0	0.052	5.0	2.2	0	0.14	4.7	0.83	0
5	0.14	0	0.030	4.3	3.6	0	0.056	5.6	2.2	0	0.14	5.7	0.84	0
4	0.20	0	0.037	4.4	3.3	0	0.072	5.4	2.2	0	0.20	4.8	0.82	0
5	0.20	0	0.041	4.7	3.4	0	0.079	6.0	2.2	0	0.20	5.9	0.83	0
4	0.14	0.5	0.022	3.3	4.0	2.6	0.043	4.4	2.6	0.8	0.14			0

In the case of conical or round shaped jets, the crater depth which develops for a clear water jet can be calculated by means of the formula (4) (see 5.1.1). This is reduced for the influence of sand by means of Figure 41 (see 5.1.2) based on the found values for c_b. The alternative method with formula (7) (see 5.1.3) is applied as well. For this the values U_b, c_b and d_b from the Tables B3 to B6 are used for the greater fall heights.
The crater diameter $2r_\infty$ is found by means of formula (6) (see 5.1.2). When shifting the pipe during discharging the crater depth will be less than calculated here. However, the crater diameter remains more or less the same.
By means of the Figures 44 and 45 (see 5.3) the width B_{min} is found for the mixture flowing across the crater edge. With this ($B_{max} = 2\pi r_\infty$ is not used) and with the formulae (13) and (14), the specific flow rate q and the specific production rate s are calculated.

The sedimentation length is subsequently found by means of Figure 48 (see 6.3) and the slope gradient by means of Figure 56a (see 8.5). The results of the extreme values of the Tables B3 to B6 are summarized in Table B7.
Based on these calculation results the possible development of the slope is indicated in Figure B2 as well as the equilibrium slope if no flow slides occur. If at a fall height of 3 m B_{max} would have been used instead of B_{min}, a slope gradient of approximately $1:8$ would have been found. It can be seen that the slope for this working method could have been considerably steeper than for bottom discharging with barges, obviously if not disturbed by flow slides.

Table B7. Crater dimensions and sedimentation length for pipeline discharging from the water surface.

magnitude	unit	fall height 19 m	fall height 11 m	fall height 3 m
crater width $2r_\infty$	m	10 – 40	8 – 30	5 – 20
B_{min}	m	12 – 18	7 – 10	2.5 – 5
specific flow rate q	m²/s	1.8– 2.4	1.4– 1.7	0.45– 1.1
specific sand production s	kg/ms	45 –200	70 –350	170 –550
sedimentation length L	m	150 –250	120 –170	50 –130
equilibrium slope $\tan \alpha$	–	$1:7$–$1:12$	$1:8$–$1:14$	$1:11$–$1:17$

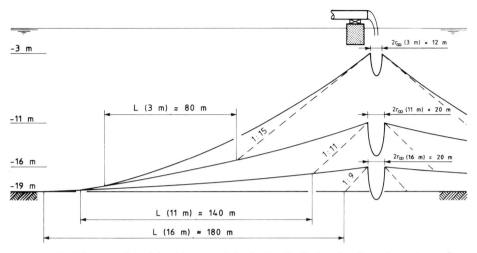

Fig. B2. Building up of the island Issungnak for hydraulic discharging from the water surface.

B4 Hydraulic discharging on abovewater fill area (Issungnak)

For hydraulic discharging on the abovewater fill area it is assumed that the sand-water mixture is discharged with the following characteristics: concentration $c_0 = 0.14$ to 0.20 and flow rate $Q = 1.5$ to 3.0 m³/s. It is further assumed that the sand has the following characteristics: $D_{50} = 225$ μm and $D_{15} = 110$ μm (no washing out of fines).

If the mixture is not spread by means of bulldozers, it would have reached the water line in channels with a width of approximately 5 m. There the concentration would sometimes have decreased, so that a variation c can be expected of 0.1 to 0.2. With these data and by means of Figure 48 (see 6.3), the sedimentation length for the underwater slope can now be determined and the equilibrium slope by means of the Figures 53a and 53b (see 7.4). The results are presented in Table B8.

Tabel B8. Sedimentation under water for hydraulic discharging from the abovewater fill area.

magnitude	unit	value
specific flow rate q	m²/s	0.3– 0.6
specific sand production s	kg/ms	80 –320
sedimentation length L	m	35 – 60
equilibrium slope tan α	–	$1:9-1:14$

From the values given in Table B8 for the resulting slope, it may be concluded that this cannot be much steeper than $1:9$ provided no flow slides occur.

B5 Hydraulic discharging just above the sea bed (Uviluk)

During the construction of this island the slope gradient was mainly determined by the construction of the underwater bunds. With the theory presented in this manual an estimate can be given of the slopes of these underwater bunds as a function of the grain characteristics and the working method. The sand is placed under water by emptying hopper dredgers via the suction pipe by means pumping.

In the following results are presented of some calculations which show that for the chosen work method indeed slopes could be expected of $1:5$.

For the emptying of the hopper dredgers by means of pumping, it is assumed that the pipe diameter was 0.9 to 1.2 m, the concentration c_0 at the point of discharge 0.14 to 0.20 and the mixture velocity U_0 4 to 5 m/s. It is further assumed that the sand had the following characteristics: $D_{50} = 320$ μm and $D_{15} = 160$ μm. The pipeline nozzle is kept all the time just above the sea bed, so that characteristics related herewith may be used as input for the calculation of the crater diameter.

The crater depth y_s corresponding with a clear water jet is calculated with formula (4) (see 5.5.1). This is reduced for the influence of sand by means of Figure 41 (see 5.1.2). The alternative method with formula (7) (see 5.1.3) is applied as well. With this, the values U_b, c_b and d_b of the Tables B3 to B6 are used for the greater fall heights.

The crater diameter $2r_\infty$ is found by means of formula (6) (see 5.1.2). When shifting the pipe during discharging the crater depth will be less than calculated here. However the crater diameter remains more or less the same.

The width B_{max} follows from formula (10a) (see 5.3) and the width B_{min} of the mixture flowing across the crater edge, is found by means of the Figures 44 and 45 (see 5.3). Subsequently the specific flow rate q and the specific production rate s are calculated both for B_{min} and for B_{max} with the formulae (13) and (14).

Finally the sedimentation length is found by means of Figure 48 (see 6.3) and the slope gradient by means of Figure 56a (see 8.5). The results are summarized in Table B9.

Tabel B9. Crater dimensions and sedimentation length for pipeline discharging just above the sea bed.

magnitude	unit	B_{min}	B_{max}
crater width $2r_\infty$	m	7 – 25	7 – 25
spreading width B	m	3 – 6	22 – 80
specific flow rate q	m²/s	0.6– 1.3	0.03– 0.26
specific sand production s	kg/ms	200 –700	11 –140
sedimentation length L	m	50 –100	3 – 20
equilibrium slope tan α	–	$1:8$–$1:13$	$1:4$–$1:7$

Based on these calculation results the possible building up of the slope is indicated in Figure B3 as well as the equilibrium slope if no flow slides occur. The slope gradient based on B_{max} appears to agree rather well with the actual constructed slope.

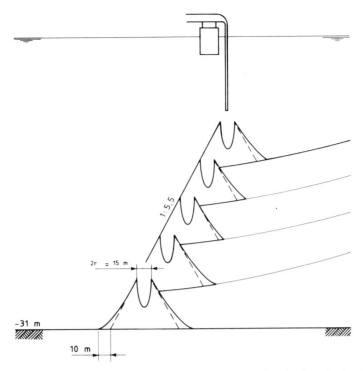

Fig. B3. Building up of underwater bunds of the island Uviluk for pipeline discharging just above the sea bed.

B6 Flow slides

In 8.4 a method is described to determine whether flow slides dominate the slope formation. This would be the case if L^* is smaller than 1, N^* greater than 1 and H^* smaller than 1. Based on the values found for the sedimentation lengths and the slope gradients in B2, it may be concluded that L^* will be greater than 1 as long as the sand body is not built up higher than 5 to 10 m. Hence during bottom discharging (up to approximately 8 m above the sea bed), flow slides will not strongly dominate construction. As a result of bottom discharging probably N^* is smaller than 1 (see 9.5.3). For the Issungnak case, for that reason, the sand will not be sensitive to liquefaction up to a level of approximately 10 m below the water surface.

However, above this level N^* will be greater than the sedimentation lengths and slope gradients found in B3 and B4, it may be concluded however that L^* is greater than 1 as long as the sand body is not built up 5 to 10 m above the already constructed level, hence near the water surface. According to formula (17) for Issungnak $h_{cr} = 6$ m, therefore H^* is smaller than 1 up to more than 6 m above this level. Consequently for levels close to the water surface flow slides will not dominate the slope formation. Beyond this level this would be indeed the case. However, this will not be very dominant.

For the steeply constructed sand bunds at Uviluk it always applies that N^* is greater than 1. From the sedimentation length and slope gradient calculated in Chapter 5, it follows that L^* is greater than 1 up to approximately 5 m above the bottom. According to formula (17), h_{cr} would be 11 m. Therefore it could be expected that for Uviluk flow slides will dominate the slope formation as soon as the elevation of the sand body is higher than 10 m above the bottom. This was not the case. Probably the value of h_{cr} is still greater for the rather coarse sand used.

B7 Conclusions

The results of the calculations for bottom discharging and pipeline placement under water is very sensitive to the spreading width B of the sand-water mixture during over-flowing of the crater edge and the spreading width thereafter (see 5.3). When applying B_{min} at the top of the slope, considerably smaller gradients are found than observed. When applying B_{max}, always a reasonable agreement is found.

According to calculations the flattest slopes result with the construction of underwater slopes, such as carried out for Issungnak, for bottom discharging with split barges. Considerably steeper slopes result for hydraulic discharging from the water surface and for hydraulic discharging on an abovewater fill area.

The fact that for Uviluk the slopes are steeper than those realized for Issungnak, besides being due to coarser sand, is mainly due to the working method of hydraulic discharging just above the sea bed. However for this working method, the risk that flow slides will dominate the process of slope formation is rather large as soon as the sand slope is built up above a certain critical value. According to the theory presented in this manual the critical height would be more than 10 m. Because of the fact that no flow slides were observed, it may probably be concluded that this critical height was greater.

However, slope gradients as realized for Uviluk ($\tan \alpha = 1 : 5.5$) seem to be the ultimate of what can be realized and only feasible for rather coarse sand with a small silt content. This is confirmed from experience with the island at Nerlerk where large flow slides occurred after initially – due to hydraulic placement just above the sea bed – a slope of $1 : 5$ was realized with sand with a $D_{50} = 220$ to 260 µm and 5 to 10 % silt or clay [34]. For Uviluk this fine material could wash away during the hopper dredge process; this was not the case for Nerlerk because sand was placed directly with a suction dredger.

SAND TUBES

C1 Introduction

A sand tube is defined as a elongated bag filled with sand. Sand tubes are used with such structures as quay wall, breakwaters, bunds etc. and are especially applied in those areas where sand is cheap and rock very expensive.

Sand tubes are also used for the construction of bunds under water to prevent sand losses during placement.

Sand tubes can have moderate dimensions such as for burlap sand bags but can be up to 6 m thick with lengths up to 60 m when using geotextile fabrics [60]. This Appendix is based on the use of larger dimensioned tubes.

The working method for placement of sand in a structure is described in Chapter 17. In the following paragraphs the specific problems of fabricating and placing of sand tubes are described. For additional information reference is made to the literature [61, 62, 63, 64, 65].

C2 Fabrication of sand tubes

Fabric and choice of seams

The fabric must be relatively strong and sand tight and resistant against damage and UV radiation. Suitable for use are synthetic fibres of polypropylene, polyethylene, nylon and polyester. In particular the first two materials are at best resistant against damage and UV radiation.

The sewed seams can create a problem because they are often weaker than the fabric itself. The welded seams require a great investment in equipment but usually are at least as strong as the fabric itself.

Filling

The filling of the tubes can be done according to Figure C1. The in and outflow nozzles must be placed as high as possible in order to achieve a proper sand filling in respectively the upstream and downstream end of the tube. The formation of an erosion hole in the upstream end of the tube can be prevented by making this erosion hole to occur in front of the tube. This situation is obtained by the application of a widened filling nozzle. The current velocity of the inflowing mixture has to be regulated properly in such a way that first the downstream end of the tube (at a slightly higher velocity than the minimum velocity) is filled. By even reduction of the velocity to the minimum velocity, continuous filling takes place in a forward direction.

The transport water leaves the sand tube through the outflow nozzle and, depending on its permeability, via the fabric. If the fabric has a large permeability, the danger exists

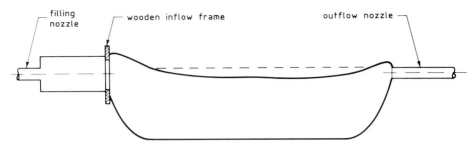

Fig. C1.   Filling of sand tubes.

that all the water has seeped through the fabric before the sand has reached the end of the tube. Then the filling process is hampered. If the mixture contains a lot of silt this may result in impermeability of the fabric.

The filling can take place on a flat surface where precautions have to be taken for support of the in and outflow nozzle (see for example Fig. C1). The filling can take place both in the dry and in the wet, in a special placement apparatus or in the bin of a (split) hopper barge. In this case the fabric is placed in the bin before filling. After filling the loose sides of the fabric are folded back across the sand and sealed.

During a proper filling process a filling ratio of approximately 90 % can be achieved at which a flattening (difference between width and height) of approximately 20 % occurs. During filling a larger mesh net develops on the sewed seam than elsewhere in the fabric, through which sand may be lost.

After filling porosities of 30 to 50 % have been measured in sand tubes.

C3   Placing of sand tubes

The method of placement of sand tubes depends on the requirements of placement accuracy. If there are no special requirements put forward, sand tubes can be discharged from the deck of a dumping vessel or from the bin of a split hopper barge. If accurate placement is required, usually a rope crane is employed for placing the sand tubes. For the support of a sand tube during placement use of a so called placing frame is necessary.

Various types of placing frames exist varying from a balance beam, from which the sand tube is suspended by means of straps, to shells in which the sand tube is laid. The discharge just above the sea bed takes place by pulling loose the suspension straps, by opening of a support or by tilting of a support. An example of a placing frame is given in Figure C2.

In order to achieve accurate joining of the tubes under water, an adequate position system is required. When the first layer is properly placed, the placement of the subsequently layers will hardly be a problem.

The stacking height is 10 to 20 % lower than the theoretical stacking height based on exact joining of round tubes.

Fig. C2.   Tube placement apparatus.

C4   Characteristics of sand tubes

Internal (in)stability
Under influence of waves and current sand can be displaced within the tube. The extent of stability depends on the filling ratio. A higher filling ratio means higher stability.

Flexibility
When sand tube(s) have to be placed on an irregular sea bed, flexibility is required to prevent damage to the sand tube. A lower filling ratio gives higher flexibility.

Friction
Depending on the material the friction coefficient between the tubes varies between 0.40 to 0.65.

Settlement
Compaction is required for stacks which are not allowed to be susceptible to large settlements. For this two possibilities are available, namely:

– dynamic compaction;
– static compaction.

The advantage of dynamic compaction compared to static compaction is, that in a short period the desired compaction is achieved in a relatively simple manner.

C5   Protection of sand tubes

There are three methods of protection of sand tubes against damage:
– external protection of the tubes with a buffer of steel, wood or concrete;
– preventing sand from departing as a result of damage of the fabric, which for example can be achieved by adding 10 % cement to the sand or by an asphalt stabilizer;
– external protection of the tubes with a coating, for example bitumen.

261

QUALITY ASSURANCE

D1 Introduction

With all kinds of production processes or rendering of services, quality has always played a major role. Before the industrial revolution it was a matter of so-called good workmanship. There was direct face to face contact between the supplier and the customer through which it was known what the client wanted and where the realization was completely in the hands of the craftsman (supplier). The emergence of mass production demanded a more scientific approach to quality and this formed the basis for the development of the present quality assurance systems. In civil engineering quality assurance is not implemented as rapidly as in other branches of industry. In the offshore and construction industry a start has been made since the beginning of the eighties. Also for other sectors of civil engineering, both social and technological developments show a growing tendency of implementating quality assurance on the manufacturing and the quality of a product.

D2 Definitions

In literature various expressions are used for quality. Main expressions are „quality", „quality control", „quality assurance" and „quality system". The corresponding definitions are formulated in international norms (BS 5750 – part I, ISO 9000–9004 and ISO 8402). The definitions are:

Quality: The degree to which the product, process or services complies with the functional requirements.

Quality control: The total of all the specific operational practices, resources and activities serving to maintain the required quality of a product, process or service.

Quality system: A documented set of activities, resources and procedures within the company organization, serving to ensure that the product, process or service meets the quality requirements of the client.

Quality assurance: The process of implementation, maintenance, review and, where necessary, improvement of the quality system, including activities proving that the quality system meets the required standards.

Quality plan: The document containing the description of the process of quality control relevant to a particular contract or project within the frame work of the (company) quality system.

D3 What is quality assurance

A product of proper quality is a product that completely meets the requirements of the client. The realization and delivery of products of proper quality is the result of proper control of all primary company activities, from market analyses or demands by the client, up to commissioning and maintenance.

Briefly and to the point: QUALITY ASSURANCE = QUALITY CONTROL

If for the manufacturing of a certain product – whether this concerns the covering of a pipeline or the making of a shaver – it is realized that this is preceded by a certain process, then quality assurance is in principle applicable. Hence the assurance of the quality of a certain project starts already in the initial stage, namely with the formulation of the functional requirements and passes through the complete process of design, work preparation, construction up to installation or commissioning, mainte-nance period and evaluation after completion.

D4 Why is quality assurance so important

Possibly, this can at best be explained by a simplified summary of the difference between the approach of a quality problem without and with quality assurance.

No quality assurance:

1. In principle only technical analysis of the problem.
2. Ad hoc actions like (extra) inspections and repairs.

Quality assurance:

1. Explicit and systematic analysis covering the technical as well as the organizational aspects before the project starts.
2. Ascertainment of instructions and communication lines from the analysis.
3. Monitor, maintain and adapt the instructions and communication lines during the execution of the project so that if a quality problem occurs it is known how to act in the quickest, cost saving way.

This is all summarized in a quality system, in a set of formulated organization regulations and procedures aiming to assure that a product, process or service meets the requirements.
Research has shown that most of the problems are organizational problems and not technical ones. Because quality assurance is a purely structural approach of how to improve the management organization the chance that the same problem appears twice is smaller with quality assurance. Apart from the general advantages of quality assurance, the implementation of a quality system is of such importance to a contractor

because the responsibility for quality is placed more and more with the designers and managers of a project. Quality assurance can even form part of a contract agreement. Besides the factor of money, more often the expected quality plays an important role with clients when prequalifying companies for a certain project.

Often companies hesitate to set up and implement a quality system because they believe that this will only cost money without gaining anything. However, it appears that despite the fact that the implementation and maintenance of a quality system costs money, lack of quality will cost even more because of all the repairs etc. which need to be done to satisfy the client in case of an inadequate product.

Quality costs are these costs which in one way or another can be accounted to quality control within the company. Also because these costs often occur in a hidden form, they are generally heavily underestimated. These costs can be divided into:

- prevention costs;
- appraisal costs;
- failure costs.

Apart from the contractor's point of view there may be also additional costs for the client when a product appears to be inadequate after the maintenance period.

Often failure costs can be high compared to the profit margin. If competitors succeed in reducing the failure costs one is quickly priced out of the market. Proper quality control goes together with proper cost control. In Figure D1 the relation is given between the

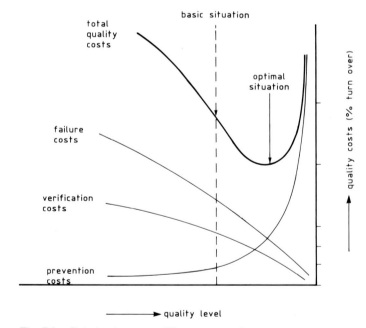

Fig. D1.   Relation between different types of costs and quality level.

different costs. It can generally be stated that working according to a quality system results in lower quality costs.

D5   How can quality assurance be implemented in a company

First of all an effective quality system is required. As explained before this is basically nothing other than a properly managed organization where targets, responsibilities and powers are uniformly documented by means of procedures and work regulations. Providing quality is therefore the result of proper management. The formulation (and availability) of procedures is not only important for the management of activities which can be accounted to the quality system, but will also stimulate the communication within the organization. It is necessary to lay down the quality system in writing and keeping it up to date in, for instance, a quality manual. It is important to identify which person (WHO) is responsible for what (WHAT), how the person in question carries out controls and verifications (HOW) and in what way this is registered and documented.

D6   Project oriented quality control

For each independent project the functioning of the quality system is analyzed and verified with the specific project, the relevant sections are subsequently used in the quality plan. The quality plan describes how the work is organized and achieved and by what means, and how the quality is controlled. During the progress of the project, the quality plan is continuously further developed and adapted.

According to the ISO 9000 norms, quality plans have to include the following items:

a. the quality objectives which are to be achieved;
b. the specific allocation of responsibilities and powers during the different stages of the project;
c. the specific procedures, methods and work instructions which are to be applied;
d. appropriate programs for testing, verification, research and audits during the relevant stages;
e. a method for coping with changes or adaptions of a quality plan during the progress of projects;
f. other measures necessary to achieve the objectives.

A quality plan can be divided into a section in which the general set up and basic assumptions of the project are described and a section in which the specific quality control of the project and/or project portions are described.

The first section comprises the items a and b in which the organization of the project also has to be described, with an emphasis on the quality organization. The specific tasks, responsibilities and powers of quality related positions (e.g. project manager, quality manager etc.) during the various stages has to be documented. In the event that more organizations are involved, it is important that the different quality saystems are tuned to each other.

The second section describes the subsequent quality activities in procedures per project item. Where possible, reference is made to detailed procedures.

In the quality plan also the reporting to be employed during the construction, must be described. For this what forms are required and who is responsible must be documented.

Hence the quality plan must provide a description of:

- all activities which influence the final product (WHAT);
- all measures which have to be taken to control the quality of each activity (HOW);
- the person responsible for a certain activity (WHO);
- how often and in what sequence (WHEN), which inspection and/or verification documents (WHERE) monitor the action.

Each partial process, which is recognized, must be controlled: quality control. This can be met by making use of regular verifications according to a check list formulated beforehand. This is known as the quality control loop (QC-loop) (see Fig. D2).

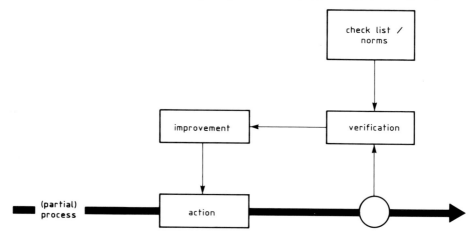

Fig. D2. Quality control loop.

The QC-loop comprises 5 steps:

1. identification of contract specifications, target values, procedures and working methods including acceptable tolerances;
2. the measurement of actual values and results; the subsequent comparison of the observations with the specifications;
3. if required rejection of the course of the process or the process result, analysis of defects and their cause and take corrective action;
4. registration of inspection and test results and corrective actions;
5. continuous efforts to improve processes and their results.

As mentioned in the first step, the QC–loop starts with the identification of the quality requirements. In practice, the ascertainment of the formulated needs takes place in

266

many forms: in writing, with drawings, verbally. As described in the QC-loop, it appears that correction takes place afterwards; when an unwanted defect in the process or product has already taken place. Therefore, such a corrective action often results in extra costs and possibly the quality of the product is permanently influenced in a negative sense. If quality assurance has to have maximum effect the process will have to be analyzed carefully beforehand and verified with the formulated needs and the risks during the process have to be assessed. For this a reliability analysis is required (see D7). The total production process has to be detailed as far as possible. The concept of the quality control loop must be applied to each individual partial process, both on intermediate results as well as on the final result. In this manner it is possible to control the complete process and to assure the quality of the product.

When working with quality assurance more is documented than in the traditional way. Although one has to pay attention not to end up in a bureaucratic mass of paper, care must be taken that all relevant data, especially appointments, conducted verifications and tests etc. must be properly documented.

D7 Reliability analysis

The aim of a reliability analysis is to minimize the risks during and/or as a result of the different project stages such as design, construction and maintenance. During the design and choice of the working method or, for example, selection of a supplier, risks during the construction and maintenance period have to be seriously accounted for. An important tool for a reliability analysis is the fault-tree. It necessitates formulation of all relevant elements including a weighting factor and a reliability margin for each individual element. In order to judge the ultimate reliability of a structure or a working method (or parts thereof), probabilistic calculation methods, amongst others, have been developed during the recent years. Based on a reliability analysis, such measures can be taken to ensure that the process or the production will have the required quality beforehand. However, despite careful preparation, there will always be factors which cannot be controlled completely. Verification and the possibility of correction afterward are always necessary, but the chance on the occurrence of defects is reduced.

APPENDIX E

REFERENCES

1. KONING, J. DE, Developments in hydraulic deep dredging. IVth Latin American Dredging Congress, Mexico City, April 1981.
2. HEEZEN, F. T. and A. C. M. VAN DER STAP, An engineering approach to under water dumped sandbodies. International Symposium on Modelling Soil-Water-Structure Interactions, Delft, 1988. Balkema, Rotterdam.
3. FISHER et al, Mixing in inland and coastal waters. Academic Press, London, 1979.
4. DELVIGNE, G. A. L., Round buoyant jet with three dimensional trajectory in ambient flow. 18th Congress of the International Association for Hydraulic Research, Cagliary, September 1979.
5. BREUSERS, H. N. C. and A. J. RAUDKIVI, Scouring; hydraulic structures design manual. International Association for Hydraulic Research, Delft, 1991. Balkema, Rotterdam.
6. BREUSERS, H. N. C., Gedrag van zandwaterstroming; ontgronding en invloed dwarsstroom (Behaviour of a sand-water mixture flow; scouring and influence cross-current). Delft Hydraulics, R1116-05, Delft, September 1985 (in Dutch).
7. RAJARATNAM, N., Turbulent jets. Development in water science, no. 5, Elsevier, Amsterdam, 1976.
8. RAJARATNAM, N., Erosion by plane turbulent jets. Journal of Hydraulic Research, no. 4, 1981.
9. BLENCH, T., Mobile bed fluviology. University of Alberta Press, 1969.
10. VISSER, P. J., J. K. VRIJLING, J. M. L. DIETEREN and P.H. POTTINGA, Breach development in dike failures. International Conference on Coastal Engineering, Delft, 1990.
11. MEULEN, T. VAN DER and J. J. VINJÉ, Three dimensional local scour in non cohesive sediments. Delft Hydraulics, Publication no. 180, Delft, November, 1977.
12. MASTBERGEN, D. R. and A. BEZUIJEN, Het storten van zand onder water, verslag experimentele vervolgstudie middelgrof zand (Sand discharging under water, report of a second experimental study on medium sand). Delft Hydraulics/Delft Geotechnics, Z 261, CO-294750, BAGT 420, Delft, 1988.
13. MASTBERGEN, D. R., A. BEZUIJEN and J. C. WINTERWERP, On the construction of sand fill dams, Part I and II. International Symposium on Modelling Soil-Water-Structure Interactions, Delft, 1988. Balkema, Rotterdam.
14. MASTBERGEN, D. R. and A. BEZUIJEN, Het storten van zand onder water, verslag experimentele studie (Sand discharging under water, report on experimental study). Delft Hydraulics/ Delft Geotechnics, Z 216/J 376, CO-284462/45, BAGT 417, Delft, 1988 (in Dutch).
15. MASTBERGEN, D. R., Zandwatermengselstromingen, verslag wiskundige modelvorming terrasvormig stort (Sand-water mixture flows, report on mathematical modelling of an terrace shaped fill area). Delft Hydraulics, Z 299, BAGT 432, Delft, 1989 (in Dutch).
16. MASTBERGEN, D. R. and J. C. WINTERWERP, Het gedrag van zandwatermengselstromingen boven water, verslag experimentele vervolgstudie (The behaviour of sand-water mixture flows, report on a second experimental study). Delft Hydraulics, Z46-02, Delft, 1987 (in Dutch).
17. DELVER, G. and H. VERWOERT, Onderzoek naar zandwatermengselstromen bij zandsluitingen, deel 1 en 2 (Research on sand-water mixture flows during sand closures). Technical University Delft, 1986 (in Dutch).
18. WINTERWERP, J. C., M. B. DE GROOT, D. R. MASTBERGEN and H. VERWOERT, Hyperconcentrated sand-water mixture flow over a flat bed. Journal of Hydraulic Engineering, no. 1, January 1990.
19. MEIJER, E. V., Experimenteel onderzoek naar mengselsprongen (Research investigation on mixture jumps). Technical University Delft, Civil Engineering Division, 1987 (in Dutch).

20. GROOT, M. B. DE, F. T. HEEZEN, D. R. MASTBERGEN and H. STEFESS, Slopes and densities of hydraulically placed sands. Hydraulic Fill Structures, ASCE Geotechnical Division Specialty Conference, Colorado State University, Fort Collins, 1988.

21. GROOT, M. B. DE, F. SILVIS, H. VAN ROSSUM and M. J. KOSTER, Liquefied sand flowing over a gentle slope. Ninth European Conference on Soil Mechanics and Foundation Engineering, Dublin, 1987.

22. JORRITSMA, J, Zandsluiting tijdelijke toegang Europoort, verslag van een experimentele studie over onderwatertaluds (Sand closure temporary access to Europort, report on an experimental study on under water slopes). Delft Hydraulics, M1118, Delft, 1973 (in Dutch).

23. HEEZEN, F. T., Taludhellingen bij zandstorten onder water (Slope gradient during sand discharging under water). Technical University Delft, 1987 (in Dutch).

24. RIJN, L. C. VAN, Mathematical modelling of morphological processes in the case of suspended sediment transport. Delft Hydraulics Communication, no. 382, Delft, 1987.

25. LINDENBERG, J., Inventarisatie adviespraktijk zettingsvloeiingen (Inventarisation practical advice on flow slides). Delft Geotechnics, CO-416509/1, Delft, 1985 (in Dutch).

26. SILVIS, F., Doorgronden wij zettingsvloeiingen? (Do we have proper insight in flow slides?). Open Discussion Day, KIvI, Section Soilmechanics and Foundations Technology, 1986 (in Dutch).

27. SILVIS, F., Zettingsvloeiingen tijdens het opspuiten van een onderwatertalud (Flow slides during pipeline placement of an underwater slope). Open Discussion Day, KIvI, Section Geotechnic, 1987 (in Dutch).

28. WILDEROM, M. H., Resultaten van het vooroeveronderzoek langs de Zeeuwse stromen (Results of lower reach embankment sections along the Zeeland waters). Rijkswaterstaat, Direction Water-management and Hydraulics, Section Engineering, Vlissingen, 1979 (in Dutch).

29. Nota stabiliteit rand bodembescherming (Report on stability of bottem protection border). SVKO, Rijkswaterstaat, Deltadienst, 22 RABO-N-82009, 1982 (in Dutch).

30. DAVIS, P. J. G, Problematiek van de randen van de bodembescherming (Issues of bottom protection borders). Polytechnisch Tijdschrift, Civiele Techniek, no. 5, 1983 (in Dutch).

31. SILVIS, F., Statistische analyse zettingsvloeiingen (Statistical analyses of flow slides). Delft Geotechnics, CO-416670/12, Delft, 1985 (in Dutch).

32. KOSTER, M. J., Meetbundel zandsluiting Marollegat (Records of sand closure Marolle Gat). Rijkswaterstaat, Deltadienst, The Hague, 1986 (in Dutch).

33. SLADEN, J. A. and K. J. HEWITT, Influence of placement method on the in situ density of hydraulic sand fills. Canadian Geotechnical Journal, volume 26, 1989.

34. MITCHEL, D. E., A lesson in hydraulic fill placement in the Arctic. 1984.

35. HODGE, W. E., Construction method for improving under water sand fills. ASCE Geotechnical Division Specialty Conference, Colorado State University, Fort Collins, 1988.

36. WINTERWERP, J. C., Zandwatermengselstromingen, verslag literatuurstudie (Sand-water mixture flows, report on literature survey). Delft Hydraulics, Z65-10, M2081-10, Delft, 1986 (in Dutch).

37. MOLENKAMP, F., Elasto plastic double hardening model MONOT. Delft Geotechnics, CO-218595, Delft, 1983.

38. JAWORSKI, W. E., Methods and control for deep densification. Symposium on Foundation Aspects of Coastal Structures, Proceedings volume 2, Delft, 1987.

39. OELCKERS, G., L. W. A. VAN DEN ELZEN and P. THOMSON, Diepteverdichting door Vibro flotation (Deep compaction by Vibro flotation). Polytechnisch Tijdschrift, Bouwkunde, wegen en waterbouw, no. 7, 1981 (in Dutch).

40. Foundation of the Eastern Scheldt storm surge barrier. LGM- mededelingen, no. 94, Delft Geotechnics, Delft, 1986.

41. Evaluatie nota verdichten (Evaluation report on compaction). Dosbouw-Deltadienst, Rijkswaterstaat, Deltadienst archives, 27VERD-N-83001, 27VERD-N-83002 en 27VERD-N-83003, Middelburg, 1983 (in Dutch).

42. PLADET, A. A., Densification of the subsoil in field practice – results obtained with a deep compaction method. Symposium on Foundation Aspects of Coastal Structures, Proceedings volume 2, Delft, 1978.

43. STAM, F., P. G. J. DAVIS and J. A. BURG, Mytilus, a floating pontoon for soil compaction for the foundation of the Oosterschelde storm surge barrier. International Conference on Compaction, Parijs, April 1980, Ed. Anciens ENPC, Paris, 1980.

44. DOUWES DEKKER, D. M. and P. G. J. DAVIS, Large-scale compaction tests for the foundation of a storm surge barrier in the Oosterschelde Estuary. International Conference on Compaction, Paris, April 1980, Ed. Anciens ENPC, Paris, 1980.

45. MÉNARD L., Application de la consolidation dynamique à l'amelioration des sols de fondation des ouvrages maritime. Internationaal Havencongres, Koninklijke Vlaamse Ingenieurs Vereniging, Antwerpen, 1974.

46. ROSSUM, H. VAN, P. STRUIK and L. VOOGT, Zandsluitingen, State of the Art (Sand closures, State of the Art). Rijkswaterstaat, BCZ-88-20.002, 1988 (in Dutch).

47. KOSTER, M. J., Evaluatie metingen op het stort Speelmansplaten I (Evaluation of measurements on fill area of the Speelmansplaten I). Rijkswaterstaat, Deltadienst, nota DDWT-85.002, 1985 (in Dutch).

48. Density measurement in situ and critical density. Delft Geotechnics, 1984.

49. NIEUWENHUIS, J. K. and F. P. SMITS, The development of a nuclear density probe in a cone penetrometer. Proceedings 2nd European Symposium Penetrometer Testing, Amsterdam, 1982.

50. LINDENBERG, J. and H. L. KONING, Critical density of sand. Géotechnique, no. 2, 1981.

51. SILVIS, F., Zandsluiting Slaak, meetverslag onderwaterstort (Sand closure Slaak, measurement report under water fill area). Delft Geotechnics, Co-285230/18 and Rijkswaterstaat, sub-report VII of the report „Zandsluitingen, State of the Art" („Sand closures, State of the Art"), BCZ-88-20.002, 1988.

52. GROOT, M. B. DE, F. SILVIS, H. VAN ROSSUM and M.J. KOSTER, Liquefied sand flowing over a gentle slope. Proceedings 9th European Conference on Soil Mechanics and Foundation Engineering, Dublin, 1987.

53. KONING J. DE, Neue Erkenntnisse beim Gewinnen und Transport von Sand im Spülprojekt Venserpolder. Verein Deutscher Ingenieurs, VDI Tagung „Bauen in Ausland", Hamburg, 1970.

54. TRAA, M. VAN, K. DE RUITER, S. BOER and R. WATSON, The use of a precisely applied sand mixture as a method for thermally insulating subsea flowlines. OTC 6156, Houston, Texas, 1989.

55. GRIFFIOEN, A. and R. VAN DER VEEN, Ontwikkeling en onderzoek ten behoeve van tunnel-fundatie door middel van onderstromen (Development and research for tunnel foundation by under-flowing). Bouw en waterbouwkunde, September 5, 1972 (in Dutch).

56. TONGEREN, H. VAN, The foundation of immersed tunnels. Delta Tunneling Symposium, Amsterdam, November 1978.

57. VOLBEDA, J. H. and G. L. M. VAN DER SCHRIECK, Trench siltation: origin, consequences and how to cope with it. Immersed Tunnel Techniques Conference, Manchester, April 1989.

58. ZWEMMER, D. and J. VAN 'T HOFF, Spending beach breakwater at Saldanha Bay. 18th Coastal Engineering Conference, Cape Town, 1982.

59. CUR-rapport 130, Manual on artificial beach nourishment. Rijkswaterstaat, Delft Hydraulics and CUR, Gouda, August 1987.

60. VELDHUIJZEN VAN ZANTEN, R., Geotextiles and geomembranes in civil engineering. Balkema, Rotterdam/Boston, 1986.

61. JACKSON, L. A., Evaluation of sandfilled geotextile groynes constructed on the Gold Coast, Australia. 8th Australian Conference Coastal and Ocean Engineering, Launceston, 1987.

62. JACOBSON, P. R. and A. H. NIELSEN, Some experiments with sandfilled flexible tubes. Proceedings 12th Conference Coastal Engineering, 1970.

63. LIU, G. S., Design criteria of sand sausages for beach defences. XIX Congress IAHR, subject B(b), paper 6, New Delhi, 1981.

64. NEDERLOF C., Verslag proef met 200 m zandworst damvak Slaak (Report on test with 200 m sand tube dam section Slaak). Rijkswaterstaat, Directorate Sluizen en Stuwen, bouwbureau Eastern Scheldt Storm Surge Barrier, ONW-R-85075, 1985 (in Dutch).
65. NEDERLOF C., Verslag proeven met zandworsten (Report on tests with sand tubes). Rijkswaterstaat, Directorate Sluizen en Stuwen, bouwbureau Eastern Scheldt Storm Surge Barrier, ONW-R-86007, 1986 (in Dutch).
66. Steady flow of groundwater towards wells. Commissie van Hydrologisch Onderzoek, TNO reports and notices, no. 10, Delft, 1964.

KEY-WORDS

For the benefit of the user the presented subjects in this manual are indicated in this list of key-words with reference to the relevant section.

key-words	**Chapter/Paragraph/ Appendix**
abovewater fill area	2.1 2.5 Chap. 7
backfill	Chap. 15 Chap. 20
bed-form types on hydraulic sand fill	2.5 2.6 7.3 8.3
berm construction	20.2 20.10
borings	3.2
– droptool boring	
– vibration boring	
– wash boring	
borrow pit	2.2
boxcut	9.5 16.8
bridge formation	17.5
building up with sand	15.6
bulking	18.3 App. A
channel:	
see bed-form types on hydraulic sand fill	
channel formation:	
see bed-form types on hydraulic sand fill	
channel width	5.3
compaction	9.5
see also compaction methods	
compaction methods	Chap. 10
concentration	App. A
– apparent concentration	
– transport concentration	
– volume concentration	4.2 7.4
concentration distribution	7.2
concentration measurement	11.2 11.5
construction types of building with sand	Chap. 15

APPENDIX G

QUESTIONNAIRE

**to obtain additional information
for the construction of sand fills**

Expanding the know-how on the construction of sand fills, depends very much on additional observations and registrations, especially on prototype scale. During the construction of a project often a number of subjects are measured to monitor the progress of the works. Possibly this manual gives cause to obtain a better picture of certain phenomena during the construction by carrying out specific measurements. We would appreciate to have this information made available to us in order to include this in due time in a supplementary edition of this manual. To provide a guide line on what data to collect, a checklist is presented below with primarily required data. Also the case histories described in the Chapters 20 and 21 provide suggestions for data to be collected. Your cooperation will be highly appreciated.

Check list of subjects

Brief description:
This describes briefly the most important functional needs of a project and provides at the same time a description of the situation.

External conditions:
Tides, current velocities, wave and wind climate, air and water temperature (only relevant in extreme circumstances) and subsoil conditions of the construction site and the borrow area (D_{50}, preferably grain size distribution curve).

Structure to be made/working method:
A description of the structure to be made and the manner in which the project is carried out.

Quality assurance:
Followed procedures etc.

Construction aspects:
Production, progress, grain size (D_{50}, even better grain size distribution curve), concentration c in the pipe line and on the fill area, crater dimensions, slope gradients (under and above water), densities.

278

A description of the measuring methods of the above mentioned parameters is also essential.

These data can be sent to: CUR
on behalf of project C 56
P.O. Box 420
2800 AK Gouda
the Netherlands

DATE DUE